P9-DFZ-497

SHAKE THEM 'SIMMONS DOWN

ALSO BY JANET LEMBKE

River Time

Looking for Eagles

Dangerous Birds

Skinny Dipping

SHAKE THEM 'SIMMONS DOWN

■

JANET LEMBKE

Drawings by Joe Nutt

Lyons & Burford, Publishers

Several pieces have appeared, sometimes in slightly different form, in these publications:

"The Apple Ancestor," Florida Quarterly, Vol. 6, No. 1.

"The Fortunes of Sassafras," Wildlife in North Carolina, Vol. 59, No. 7.

"Shake Them 'Simmons Down," Wildlife in North Carolina, Vol. 57, No. 12.

"The Tree That Came Out of the Shade," Wildlife in North Carolina.

"Tree-Knowledge," The Hiram Poetry Review, No. 16.

"The Woman Becomes a Dryad," Inlet, No. 12.

Printed in the United States of America

10 9 8 7 6 5 4 3 2 1

Design by Kathy Kikkert

Typesetting and composition by Sam Sheng, CompuDesign

Library of Congress Cataloging-in-Publication-Data

Lembke, Janet.

Shake them 'Simmons Down and other adventures in the lives of trees / Janet Lembke; drawings by Joe Nutt.

p. cm

ISBN 1-55821-350-3

1. Trees—United States. 2. Trees—United States—History. 3. Lembke, Janet. I. Title.

QK115.L38 1996

582.160973—dc20 96-14375

CIP

For Mo Wixon,

the neighbor who says,

"Planting a seed,

you can't deny the hope that goes with it."

And in memory of Hannah,

1960–1995,

the daughter who could cast seeds on a stone

and watch them grow.

Contents

Acknowledgments · ix

Out on a Limb: An Introduction · 1

Dryads in the New World: Black Tupelo · 13

The Woman Becomes a Dryad · 25

The Tree That Came Out of the Shade: Loblolly Pine · 27

Shake Them 'Simmons Down: Persimmon · 43

Eat Them 'Simmons Up · 49

In Search of Harmony: Yellow Poplar · 53

Tree-Knowledge · 66

Sweet Gum, with Occasional Birds · 67

The Fortunes of Sassafras · 85

The Flavors of Sassafras · 98

Hogoo: An Interlude in the Back Field · 101

Mr. Jefferson's Nut Tree: Pecan · 117

Pecan: The Inseparable Adjunct · 127

The Apple Ancestor · 129

Where, Oh, Where Is Pretty Little Susie?: Pawpaw · 131

From the Pawpaw Patch to Little Susie's Kitchen · 150

Suds: The Yuccas · 153

Erasmus Darwin and the Upas Tree: An Interlude in Other Places, Other Times · 165

Daphne · 179

The Devil's Walking Stick · 181

A History of Obsolescence: Osage Orange · 193

A Good Truck-Fixin' Tree: The Catalpas · 203

Life Stories: The Sumacs · 213

Books to Leaf Through: Further Reading · 227

Acknowledgments

Many people have generously contributed ideas and information to my adventures in the woods—and spent countless hours putting me back on the right path when I went astray. Some are fully named in the stories that follow. Others who deserve an equal share of light are listed below:

- Harry Haney, Jr., and Mark Hollberg, foresters, whose delight in the green world is catching
- John Herington, good friend and perfect host, who gleefully introduced me to Erasmus Darwin
- Jeannie Kraus, botanist, who has a glorious talent for coming up with the scientific and common names of plants the instant anyone asks

Nor should my patient husband, the Chief, and my always encouraging editor, Peter Burford, be cast into the shade. To all of you, and to any whom I may have overlooked, my happy, heartfelt thanks.

OUT ON A LIMB

An Introduction

■

Let no one visit America without first having studied botany; it is an amusement, as a clever friend of mine once told me, that helps one wonderfully up and down hill, and must be superlatively valuable in America, both from the plentiful lack of other amusements, and the plentiful material for enjoyment in this. . . .

—Mrs. Frances Trollope
Domestic Manners of the Americans, XXVII

Ever since human beings stood upright somewhere in east Africa and began to walk upon this earth, we've associated ourselves with trees in many ways. Some of the associations we've come up with are matter-of-course: trees make war clubs and serve as fences, they furnish wood for bows and dwellings, they fuel the fires that keep us warm, they shade us beneath cool green leaves, they feed us with succulent fruit, they delight our eyes and stand witness to our histories. Other associations—in the Western world at any rate—are matters of choice, culture, and perception. Our ways of looking at trees seem to be at least as multitudinous and multifarious as trees themselves. Consider such things as the king of the wood, the golden bough and the golden apples of the Hesperides, and the sacred ash tree that is reported by Norse myth to support the universe. Consider also tree houses, family trees, and tree hugging, not to mention putting down roots or going out on a limb.

Trees, of course, have been around a lot longer than we have. According to the Book of Genesis, the world has been green with grass and other herbage and trees that enclose their seed in fruit ever since the third day of creation, while we weren't molded into being until day six. Scientists (lacking precise tables for converting the length of the divine day into mortal hours) have devised their own timeline.

For grass, herbage, and trees, it goes something like this: six hundred million years ago, just after sunrise on the third day, the world was at last wrapped in enough atmospheric oxygen to support life on land. And by midmorning, about four hundred million years ago in the geologic period called the Late Silurian, the earliest land plants—things like the club mosses and horsetails—took advantage of the situation. The first seed-bearing plants appeared some fifty million years later, in the Late Devonian period; among them were the aboriginal cycads and conifers, the gymnosperms or "naked seeds." (Let there be redwoods and ginkgoes.) The flowering plants, the angiosperms or "contained seeds" with their protective sheaths, took root during the afternoon of the third day, 120 million years ago in the Early Cretaceous period. (Let there be magnolias, yellow poplars, sweet gums, sassafras, oaks, and maples.) Trees had long attained a firm hold on earth by the dawn of the sixth day, an event that took place a mere four million years ago, when protohumanity first appeared on the earth. In the flash of time since then, we have pushed through many woods, felled many forests, and busied ourselves by fashioning trees into totem poles, canoes, and houses. And we've named every tree we've come across, although, as the poet W. S. Merwin has most truly remarked, "their names have never touched them."

The naming of trees is a curious matter. Their common names delight imagination—tacamahac, kingnut, possum oak, horse sugar, sugarberry, Christmas-berry, candleberry, farkleberry, bodark, stinkbush, Indian banana, shadblow, tulip tree, Judas tree, tooth-ache-tree, devil's walking stick, old man's beard, and a hundred hundred others. The common names also preserve native languages, from the Malaysian upas to the North American tupelo, catalpa, sassafras, and persimmon. But most of the scientific names for trees and for plants as a whole are so plain, so stolid, so resolutely unfanciful, they seem impoverished. The animals fare better. While many of their binomials are also nothing more nor less than straightforward descriptions, phrased in Latin, quite a few dare to tell stories. Designations for birds often brim over with narrative. The osprey's name, *Pandion haliaetus,* recalls the brutal Greek king Pandion, who was relieved by divine fiat of his human shape and turned into a fish-eating raptor; the names of two twittering birds—the purple martin, *Progne subis,* and the tree swallow, *Iridoprocne bicolor*—summon the memory of his wife, Procne, whose speech consisted only of inarticulate sounds because he had cut out her tongue. Myth also takes wing with the lepidoptera. The several species of Io moths commemorate a princess for whom Zeus lusted; young Io was turned into a heifer by Zeus's jealous wife, Hera, but Zeus prevailed after all by becoming a bull. The Polyphemus moth, *Antheraea polyphemus,* tells about the one-eyed giant who almost succeeded in preventing Odysseus's return to Ithaca from the Trojan War. The Palamedes swallowtail, *Papilio palamedes,* tells of the prince responsible for sending Odysseus to Troy in the first place; feigning madness, Odysseus hoped to stay home, but Palamedes discovered the ruse (and was later killed in an act of revenge).

Animals without wings share in the storytelling, too, which may depart from myth and take other forms. There is, for example, *Gorilla gorilla,* named not once but twice for a tribe of hairy women, the Gorillai, reported to live on the west coast of Africa; this rumor reaches the present day through a very old Greek version of a geographical account first written in Phoenician by one Hanno, a Carthaginian navigator who supposedly conducted his explorations around 500 B.C. And not only ancient lore but also human judgments of the animal in question may invest its name: *Ursus horribilis*—the dreaded bear, the bear that makes you tremble with fear—for the grizzly; *Diadophis amabilis*—the lovely, lovable banded serpent—for the Western ring-necked snake. (Who would mind meeting a snake like that?)

But when it comes to stories or even to opinions, the plants' binomials are mostly mute. Exceptions do exist—*Nyssa,* for one, the genus name for the tupelos, tells of nymphs and a bibulous god—but botanical nomenclators, unlike zoologists, appear almost invariably to have been literal-minded men with feet planted firmly on the ground, eyes focused on the immediately real and tangible. Imagination in almost any form, from tall tales to metaphor, seems to have been off-limits. Objective descriptions ruled the day. The resulting binomials, especially the species names, dwell on colors, shapes, similarities, uses, geographical locations, and other features that can be directly observed; a small handful from the overspilling cornucopia includes *aurantifolius,* golden leaved; *campanulatus,* bell shaped; *triloba,* three lobed; *liliaceus,* lilylike; *officinalis,* medicinal; *orientalis,* Eastern; *scandens,* climbing; and *serotinus,* late flowering or late ripening. It's as if the name-bestowing botanists, all thoroughly enamored of the verdant world, didn't know quite what to make of it and made the best of the situation by rejecting stories and hunches in favor of objectivity.

But the people who should be most skilled at creating the kind of fantasy that tells the truth—the poets—don't seem to know either. (This is the point at which Joyce "Only God can make a tree" Kilmer insists on being mentioned—but only in a parenthetical whisper: he's a maker of sentimental bad verse, not a poet. The sole virtue of his jingle may be that it has stirred generations of people, especially schoolchildren, to notice trees, albeit trees that are uncharacteristically maternal and devout.) My inquiries into the associations of poets with trees have involved considerable searching but scanty results; they indicate that poets attempt infrequently to make literary contact. As with the genus name for the tupelos, exceptions do occur, and Walt Whitman is prominent among them. In "The Song of the Open Road" he wonders, "Why are there trees I never walk under but large and melodious thoughts descend upon me? / (I think they hang there winter and summer on those trees and always drop fruits as I pass;). . . ." And listen to a passage from one of his "Calamus" poems:

> I saw in Louisiana a live-oak growing,
> All alone stood it and the moss hung down from the
> branches,

> Without any companion it grew there uttering joyous
> leaves of dark green,
> And its look, rude, unbending, lusty, made me think
> of myself,
> But I wonder'd how it could utter joyous leaves
> standing there alone without
> its friend near, for I knew I could not,....

"Uttering joyous leaves"—how rudely, lustily, resplendently the foliage rustles and shines! And in his "Song of the Redwood," Whitman even projects himself into the interior lives of the "majestic brothers," whose tops rise two hundred feet in the air, whose trunks wear bark a foot thick; these trees know joy and know, too, that it is their destiny to yield to the "new culminating man," "to duly fall . . . to disappear, to serve." The contacts here are strictly literary, however; the live oak is less a tree than a foil for Whitman's thoughts about his need for companionship, for love, and the redwoods, along with the trees bearing large and melodious thoughts, are important not in their own right, but for their material or psychic usefulness to humankind. Trees like Whitman's—trees as literary devices, trees formed in the poet's image—are a species found only in the habitat furnished by poems. The job of a poet, after all, is not to define the nature of trees nor even to comment on them but rather to interpret the human condition.

As for trees as trees—trees that can be seen, touched, and maybe hugged—today, many people regard them not only with admiration or outright passion but with a feistily protective zeal. Organizations working to save the trees proliferate; they've risen up around the globe like a great coppice sprouting from an extensive network of underground runners. Rainforest Action Network, Save America's Forests, Ecoforestry Institute–US, Forest Guardians, Japan Tropical Forest Action Network, Reforesting Scotland, Estonian Green Movement, Heartwood—these are a representative few of the groups that would ensure the survival of old-growth trees in the red-cockaded woodpecker country of the Southeast and in the spotted owl and marbled murrelet forests of the Northwest; that would preserve the neotropical wintering grounds of warblers, thrushes, and Monarch butterflies and preserve, as well, their more northerly summer breeding sites; that would ban erosion-causing clearcuts and halt the deforestation that occurs so that oil pipelines and highways can be built and land developed; that would keep W. H. Hudson's mansions forever green, their many rooms filled with an uncountable variety of living things, and keep Longfellow's pines and hemlocks forever murmuring, even though the forest in which they grow is no longer primeval. These groups sometimes go overboard, insisting too loudly, too belligerently, on the absolute rightness of the cause and the correctness of the information on which plans and actions are devised. (They have, thank goodness, avoided the silliness and ignorance of some of the people who espouse animal rights, including, recently, the rights of fish not to be tormented by baited hooks. I've not yet discovered any group that advocates plants' rights; though plants do experience shock, and perhaps pain as well, people will continue to excise or poison dandelions, subject roses to ruthless pruning, and hack down trees.) But efforts at conservation, to succeed, will always entail a balancing act, a compromise between the frequently at odds requirements of humankind and the rest of creation. Apples and oranges, spotted owls and loggers—where are the formulas for equivalence that allow one to be judged in the same arena as the other? The best we can do may be only to choose among the facts and theories available at the time and come up with whatever solution seems to do the least harm all around. But people who would save trees, forests, wilderness, and the diversity therein certainly accomplish one noble purpose. They serve as a two-by-four—the kind that was used to bash the legendary mule: sometimes it takes great force to gain someone's attention.

My attention has been struck by something else. As I poked around in the real woods and through a thicket of paper and print, studying the ways in which people look at trees and use them, it became apparent that more than an ocean separates the New World from the Old. A curious gulf also lies between two sets of perceptions. Though genus and even species may be identical, a tree on the western side of the Atlantic is not regarded in the same light as its counterpart on the eastern side. Many present-day Americans have their roots in Europe and Africa, but when our ancestors first arrived on this wild new shore, it seems they left behind many of their cultural resources for seeing and understanding natural things. Or forgot them. Or had them knocked topsy-turvy and inside out by a grand and spectacular landscape that was also daunting, terrifying, and so alien that old perceptions did not fit. (And to what extent did the generations-long struggle to come to terms with a strange continent form the American character?) The difference between perceptions is this: thick as moss and lichens, myth and superstition cling to Old World trees, while New World trees grow bare of such encrustations. In the Old World, birch and rowan, ash and alder, willow, fir, white poplar, hawthorn, and oak—oh, especially oak—have been immemorially sacred, whether standing alone or massed in groves. In Greece, oaks belonged to Zeus, and at Dodona his priestesses would sit beneath the mighty trees listening to the instructions that the god transmitted through the cooing of the doves perched on the holy branches. In Britain and in Britanny, the Druids—whose name is derived from a Welsh word meaning "oak seer"—delivered prophecies with the help of twigs from their dedicated tree. Not only oaks but all other trees housed spirits, divinities, dryads. And witches have ever dwelled, cackling and stoking ovens, in deep Teutonic woods, while Russian forests shelter Baba Yaga and her hut that is perched on chicken legs. And green men—not human nor arborescent plant, but nonetheless living and magical concatenations of bark and leaves—may still roam in England (one such eruption was recorded as recently as 1970 in Kingsley Amis's eponymous novel *The Green Man*). So it goes in the Old World: trees are not what they appear to be but are invested—one might say hallowed—with layer on layer of legend and belief. They're larger than they seem.

The New World's trees are simply trees. They have, of course, assumed psychic importance beyond their value as homes for spotted owls, ornaments for gracious lawns, fall tourist attractions, and providers of useful products from flooring and Christmas trees to turpentine and maple syrup. They are an evergreen source of metaphor: sturdy as an oak, graceful as a willow, stock and scion, chip off the old block. And some individual trees have been granted special regard and given proper names to note particular human occasions or honor particular people. One was the Charter Oak of Hartford, Connecticut, which served in 1687 as a hiding place for the colony's charter so that it could be kept safe from the land-grabbing British governor of New England; the tree was blown down in 1856. Another named tree, still living and perhaps four hundred years old, is the Treaty Oak of Austin, Texas, a town designated in 1839 as the capital of the Republic of Texas and in 1845 as the capital of the Union's newest state; nearly a century and a half later, in 1989, the tree was almost killed by a vandal who applied an herbicidal poison to the soil around its roots, but it was saved by more potent applications of money and scientific expertise. The most venerable of the New World's named trees is the General Sherman Tree of Sequoia National Park in California; its age is estimated at three thousand years, and standing 272 feet tall, measuring a good 113 feet in girth, it is indeed a giant amid giants. I do not know if General Sherman ever laid eyes on it, though he did work as a banker in San Francisco during a low civilian spell between the dashing and successful episodes of his military career.

But, New World and Old, it's as if the lives of the plants—from violets, dandelions, and crabgrass to willows and devil's walking sticks—are so far removed from our human lives that we can approach them only with a certain respectful caution, not to mention reverence. Flowers are somehow easier: they're smaller than we are. But trees are creatures of a far more imposing order, with lives that are probably impossible for us to comprehend. Their habits are open to understanding, yes—reproductive physiology, the mechanisms of photosynthesis, the chemistry of autumn colors, and myriad other aspects of treehood that can be quantified. The quality of life may also be partly determined from the vigor or the unthriftiness of growth. But what of communications, one tree to another? What of volition, one tree by itself or in concert with others? And what of joy or pain?

It's in our power as human beings to imagine ourselves as eagles, whales, horses, bears, cats, sharks, snakes, and even insects (note butterflies and various roaches from Archie, friend of Mehitabel the cat, to Gregor Samsa). And for these easy transformations, we choose animals that possess central nervous systems and are able to move at will. I doubt, however, that many of us can imagine ourselves as trees,

which
do
little
but
stand
there.

Become a tree? Can anyone engage in that ultimate act of sympathy?

But trees certainly attract notice (though they do so without the slightest intent or semblance thereof). They're obvious and inescapable facts of life, and therefore worth pondering. One of my friends focuses on trees—ancient olives, Druidic oaks—as an aid to meditation. And the tree outside her window at a British lying-in hospital furnished the name for her newborn daughter: Cerris, for *Quercus cerris*, the Turkey oak. Her connections with trees were formed, however, by her years in England and Greece.

The trees most familiar to me are the myth-bare trees of the New World. Many wear Indian names; most have no relatives in Europe (though I think that all have dryads, just as Greek trees do). In this arboretum of stories, I've looked at selected trees through the lenses available to someone who, for all the wishing in the world, can't be a tree: botany, of course, and history, medicine, folklore, cookery, and longstanding personal acquaintance. Weeds pop up here and there, along with recipes and poems, including excerpts from a botanical treatise composed in heroic couplets. The specimen trees of the arboretum are scattered (but thriving) up and down hill in my home states of Virginia and North Carolina, and in many other high or low venues clear across the country. The stories promote no particular environmental agenda. They simply celebrate a huge green world and acknowledge the reliance of my own life, and that of other animals, on the lives of trees. Their exhalations are the air we breathe.

Up and down hill, the trail through my woods is open, with each tree, each plant in the collection plainly labeled. The sign at the entrance reads WELCOME.

DRYADS IN THE NEW WORLD

Black Tupelo

Eye-catching, fiery, mythic—all these words apply to black tupelo trees, or black gums, as they're often called. The particular member of the species that grows in our yard near water's edge also deserves words like vexing, stubborn, secretive. And it's prodded me to think about the nature of those minor deities, the dryads, whose domain is trees.

The tree in our yard on the shore of coastal North Carolina's wide and salty river Neuse certainly commands notice. In summer, it wears a glistening helmet of dark green leaves that only nesting cardinals seem to penetrate easily. At summer's end, the leaves ignite and keep blazing well into fall. In winter, bare branches and twigs stand against pallid sky and grey water like a closely woven cage of fine black wire. A few wispy streamers of Spanish moss, caught on the wire, flutter in the wind. Five loblolly pines surround the tree and spread a canopy of green needles far above its crown. Compared to them, the black gum is a runt, but it has nonetheless succeeded in making them look thin-stretched and spindly, toothpicks still clad in bark. At only nine feet our tree's height is modest, almost timid, but every year its long, slender branches, held parallel to earth, spread another few inches beyond the trunks of the sentinel pines. The crown now measures fourteen feet. That luxuriously sprawling black gum has managed to controvert the notion that a respectable North American hardwood should grow upward, not out.

As for the vexing and secretive part, our tree long guarded its identity, as if its very life depended on silence. I examined the bark and the mode of growth, found springtime clusters of minuscule flowers, gathered samples of the glossy leaves, consulted every field guide I could find—and failed for seven years to discover the tree's name. The problem was not lack of acquaintance with the species. Black gums grow generously here at Great Neck Point.

They congregate in the hedgerows and woods; they stand singly—one here, one over there—on lawns where everything is clipped and trimmed; they rise willy-nilly in yards as unkempt as ours. And I've known what they are.

In fact, one black gum is among the Point's most special and exciting trees—exciting, that is, to anyone who watches birds. (Audubon was right to paint a blackpoll warbler flitting amid the leaves of a black gum.) I check on the tree almost every day, for it grows in the yard that's located immediately beyond our rural route mailboxes. Trunk slender and straight, branches high aloft but held parallel to earth in true black-gum style, it's a graceful tree, nearly as tall as the white oaks with which it shares well-landscaped premises. And it's a kiss-me tree: at least a dozen clumps of mistletoe cling tightly not only to the stouter branches but also to some of the outermost twigs. (Don't believe statements that mistletoe prefers oaks. That's booklore. Reality roots the American mistletoe—not the equally parasitic Old World Druidic stuff but the New World's *Phoradendron serotinum*—in many hardwoods from oaks to maples and black gums. Judging from local examples, it definitely prefers the last.) The black gum near the mailboxes is altogether an enticing tree, although I cannot say from either observation or personal experience that its abundant mistletoe has ever stimulated human osculation (I hope it has). But I can say that the tree itself summons birds and works on them an ancient September magic.

Gathered in this black gum, the birds are peaceable. For ten days on either side of the autumn solstice, aggression and intolerance are put aside. The mockingbird does not rush at every newcomer with flapping wings and a loud squawk. Holding back their usual jeers and taunts, the blue jays do not muster here to drive off other species. Brown thrashers and red-eyed vireos, usually reclusive, consort out in the open with robins, cardinals, summer tanagers, grackles, and four kinds of woodpeckers. Starlings swoop in, too, mainly young ones feathered in speckled brown, and in greater numbers than at any other time of year. And look—migrating birds seen in no other season fly in to linger for a day or a week: scarlet tanagers, Cape May warblers, rose-breasted grosbeaks, Baltimore orioles, veeries. There's room in the tree for all these birds—and all of them at the same time. There's food enough to go around.

In September, because of the birds, this particular black gum also summons people. They come in cars and vans from twenty, thirty, fifty miles away to park themselves in folding chairs brought for the occasion; with binoculars and spotting scopes, they wait to see what may appear. They're never disappointed—except, perhaps, by my failure to answer the question that always comes up: Why do the birds flock in great numbers to this one tree while they alight only by ones and twos in the other fruit-laden black gums that grow nearby? I don't know.

But I do know how this black gum and all the others of its kind call in the birds. Black gums catch fire. Their leaves undergo a sort of spontaneous combustion well before those of any other trees in the eastern United States manage to respond to the approach of fall. The botanist Julia Rogers, writing in 1905, described the phenomenon this way:

> In early fall the rambler in the woods is often startled to see on the mossy carpet in front of him a thick, shining leaf, part of which is still deep green and part as red as blood. It is the tupelo's signal that winter is on the way. Look up, my friend, and the branches above show only a few leaves colored like the one you found. Come again in a week or two and the tree is ablaze with reds of every shade. It is a pillar of fire. . . . Who can fail to know the tupelo in the glory of its dying foliage? Certainly no rational being, if he has eyes in his head, and the tree in his neighborhood.

No rational bird can fail to know it, either—neither the permanent avian residents of the neighborhood nor the migrants winging through. Birds discriminate among the foods available in any given season and show clear preferences for one bug or berry over another.

In mid-September on the Carolina coast, passing up the abundant offerings of beauty berry, Virginia creeper, and privet, they zero in on the black gum's fruit (or in the case of vireos and warblers, on the insects that the fruit attracts). Like cherry, plum, and olive, the fruit is technically a drupe—a single seed, or stone, encased in flesh. And the drupes of the black gum, growing in small clusters, are blueblack ovals no bigger than a pea. I sampled one: ugh, spit-out sour. But birds have only a rudimentary sense of taste. More important, they recognize what's good for them. The extraordinarily high fat content of black-gum drupes—as much as thirty percent—provides a premium-grade fuel for migrants and stay-at-homes alike. Because of this fat content, however, the drupes may be quicker to spoil than fruit less well endowed. So, when fall approaches, the black gum makes flaming advertisement for its ripened wares. (As is their wont, the scientists have given a name to this kind of announcement: the tongue-twisting "foliar fruit flagging.") Though birds taste little of what they eat, they do perceive colors. The black gum's glowing red leaves call them in. Nor does the tree go unrewarded; the birds give compensation for good food by spreading seeds.

These seeds that delight the birds but make me pucker are reputed to have tickled the taste buds of other animals, including the human kind. In his 1709 survey of the natural history of the Carolinas, the English surveyor John Lawson made this report:

> Of the Black Gum there grows with us two sorts; both fit for Cart-Naves. The one bears a black, well-tasted Berry, which the *Indians* mix with their Pulse and Soups, it giving 'em a pretty Flavour and scarlet Color. The Bears crop these Trees for the Berries, which they mightily covet, yet kill'd in that Season, they eat very unsavory; which must be occasion'd by this Fruit, because, at other times, when they feed on Mast, Bears-Flesh is a very well-tasted Food. The other Gum bears a Berry in shape like the other, tho' bitter and ill-tasted. This Tree (the *Indians* report) is never wounded by Lightning.

I agree that bear is indeed "well-tasted" (the meat brought my way was evidently not acquired in black-gum time). But I'm not sure that I can accept Lawson's suggestion that adding black-gum fruit to other food is the secret of making it humanly palatable. It sounds to me like a sure source of ruination.

Manufacturing Lawson's "Cart-Naves," stout wooden hubs for the wheels of carts, from the black gum might not have been easy. Its wood long had a reputation for being difficult to work. And its name, according to a 1907 article published in the trade newspaper *American Lumberman*, "has been synonymous for many years with all that is dark, crooked or unmanageable. Lumbermen whose word ordinarily would not be doubted have been heard to say they had known a freshly sawed gum board, when the sun shone upon it, to curl up and crawl out of a lumber yard almost as quickly as a black snake." The secret to manageability, however, lay in the curing. When properly dried, the wood was put to use in many ways, from flooring and sheathing to barrel staves.

But black gum is good for more than manufacturing sturdy boards; it's good for more than fueling birds and stuffing the hungry bellies of covetous bears. It might even furnish a bear with a den. As a black gum ages, it breaks; a branch cracks off, or the trunk itself is topped. With the help of insects and fungi, decay sets in. In time, the entire heart of the tree may be eaten away. Many old black gums, though quite alive, are completely hollow. As the tree grows new wood on the outside, the inner cavity gets larger. I've read of black gums fully two feet in diameter but with trunk walls a mere three inches thick; yet leaves sprout each spring, turn scarlet every fall, and the fruit continues to invite the birds. The makers of lumber can't use such empty trees, but a host of other creatures can and do. The cavities provide nest spaces for woodpeckers, squirrels, and raccoons; daytime roosts for dozing owls; and domestic articles for country folk. Sections of hollow trunk make beehives, storage bins, and rabbit traps, all known as "gums"—bee gums, flour or onion gums, rabbit gums. And the tree's twigs, well chewed, may serve as toothbrushes or snuff dippers.

Black tupelo, black gum—the assignment of color seems meant to distinguish the tree first from other kinds of tupelo and second from the sweet gum, a tree with leaves that blaze as ardently, though not so early, in the fall. Some people also call the tree sour gum—sour for the puckery drupes, and gum out of habit, for the tree produces nothing, not sap nor resin, that might be properly considered a "gum." And, once upon a long-ago time, somebody dubbed it the pepperidge tree; nobody now knows why, but this designation is still found in the field guides. It also figures in the trade name of a well-known bakery, Pepperidge Farm, that may have had its start on a country place loaded with black gums.

The other gum to which Lawson refers, the gum "never wounded by Lightning," is the water tupelo. He may not have been aware that two more species of tupelo grow in North America—the Ogeechee lime and the swamp tupelo (though the latter may be a variety of the black gum, rather than a separate species). All are trees that like to keep their feet wet; they thrive in river swamps and ponds or at their soggy edges. Like the cypress, the water tupelo swells at its base to give it stability in mucky, often water-covered soils. The black gum tolerates conditions somewhat less saturated than the others require, but it too is a water-loving plant and requires moist, poorly drained soil. Though the ground here often looks quite dry, Great Neck Point and much of the coast are especially friendly to the black gum because the water table in these parts is high. It may be this greater tolerance that has allowed the black gum to spread over much of the eastern United States, from central Florida into southern Maine and from eastern Texas into Michigan, while the other tupelos confine themselves largely to the South and also, in the case of the water tupelo, to the Mississippi River basin as far upstream as southern Illinois.

It's entirely fitting that these four trees native to North America are known to this day by a Native American name—*ito opilwa,* meaning "swamp tree," which Europeans picked up from the Creek Indians of the Southeast. The word, transformed to "tupelo," was subsequently placed on all four related species. They are in fact so closely related that they belong to the same genus. The black gum is *Nyssa sylvatica,* Nyssa of the woods; the water tupelo, *N. aquatica,* Nyssa of the water; the swamp tupelo, *N. biflora,* double-flowered Nyssa; and the fourth, *N. ogeche,* Nyssa of the Ogeechee, the river in southeastern Georgia where this species was first noted. Like many of North America's indigenous trees, the tupelos have no relatives in Europe. The other members of their genus are found in the Far East, in China, Tibet, and Malaysia. And the genus belongs to the Nyssaceae, the Sourgum family. (A few botanists assign it to the Cornaceae, the Dogwood family, perhaps because dogwoods, like the Nyssa species, often show scarlet leaves early in fall to flag down passing birds and invite them to dine on red-ripe fruit.)

Nyssa this and Nyssa that—why all these Nyssas? By a somewhat winding path, the answer leads to dryads (and to that vexing, stubborn, secretive tree in our front yard). Back in the mid-eighteenth century, when Linnaeus had taken on Adam's task of putting names to all the plants and animals then known, he himself took *Nyssa* over the other available choices. One hundred fifty years later, in 1905, Julia Rogers made ladylike complaint that Linnaeus had selected a name associated with the ancient god of wine and overindulgence: "It was the fashion for the old botanists to give new plants names derived from classical mythology, without much thought of appropriateness." But *Nyssa* is appropriate, I think, and I shall soon give a reason. And I cannot agree with her that using mythological names was the botanical fashion in Linnaean days. The zoological nomenclators certainly employed their classical educations by blessing many animals with binomials drawn from Greek and Latin myths, but the botanists, apparently more down-to-earth, tended to avoid such references in favor of strictly descriptive terms. But like every generalization, this one has exceptions. And *Nyssa* is one.

Once upon a time before time began, Nyssa was a nymph who, with her four sisters, had the job of caring for that holy terror, the god Dionysos, son of Zeus and one of his many mistresses. When Dionysos was born, Zeus's wife, Hera, threw one of her legendary tantrums; so to keep his latest by-blow safe, Zeus caused the child to be spirited away by the messenger god, Hermes. Through infancy and boyhood, Dionysos lived in a high, wild mountain cave, where the guardian nymphs fed him on honey and kept him safe from harm. Or kept him as safe as they could; while in their care, the young god discovered the art of making wine. Indeed, the idea may have come to him from regular indulgence in fermented honey, a potent ingredient of the drink called mead. And so began the reign of Dionysos. Nor did he preside simply over the products of the grape but also over fertility, for which he was adored in various ecstatic and dangerously orgiastic rites (for details, see Euripides' *Bacchae*).

Julia Rogers to the contrary, I think it entirely appropriate that trees possessed of a grand and bibulous thirst (albeit for water) are associated with such a god.

I am not so much interested in Dionysos, however, as in his nurses and their later connection with tupelo-kind. Their names, according to ancient sources, were Macris, Erato, Bacche, Bromie, and Nysa with a single *s*. Today, they might be called Beanpole, Hot Pants, Hothead, Loud Mouth, and Hobble (for Nysa means "lame"). But these individual names, which clearly, colorfully denote quirks of body or behavior, are less important than the sisters' collective identity. For their service to his son, Zeus lifted them up from their mountain and set them down among the constellations. Their soft-gleaming fires are visible today: they are the Hyades, Makers of Rain. Likely, they were water nymphs on earth before they ascended to the stars. And who more qualified than one of them to lend her name to water-loving trees?

But no water nymph guides the fortunes of these trees. Though the tupelos thrive in wet places, they do not shelter any spirit of spring or brook, bog or swamp, mist or downpouring rain, or Dionysian wine. Like an ancient Greek, I'm sure of that. I'm sure that water nymphs stay in their fluid realms. It's dryads—wood nymphs—that inhabit trees.

And of all the nymphs, only the dryads are mortal. Earth's primeval water is with us still, wheeling through a never-ending cycle of evaporation, condensation, precipitation, and back again to evaporation; in the light of human brevity, earth's mountains rise and wear away so slowly that they, too, seem meant to last forever. Their nymphs—the naiads of pond and puddle, the oceanic nereids, the oreads within granite, limestone, and basalt sill—are equally eternal. But trees die. It makes sense—ancient Greek sense, at any rate—to believe that each tree's indwelling spirit suffers the tree's death. Nor do trees necessarily succumb to old age; they fall to lightning or the chain saw. Their dryads tumble with them to the ground. But new trees sprout and grow, new dryads rise with them—a renewable resource, unless the habitat is damaged or destroyed. It makes modern sense, at least to me, to believe that the notion of dryads is not at all far-fetched—not if these ladies are understood as representing the life of the tree. More than representing it; investing each tangible, bark-clad life with something larger than that life. The dryad in every tree connects it with all that is timeless and holy.

Except, I'm not sure that all dryads are ladies. Old and persistent rumors have it that some of them dallied often with satyrs, but that's not what I mean. No implication of unladylike behavior is intended. What I mean is, I'm not sure they're female. Or, not all of them.

I've circled back to the vexing, stubborn, secretive, enchanting black gum that grows in our front yard near the river. That tree is male. Beyond all disputation, it is male. *N. sylvatica* exhibits a set of reproductive arrangements that scientists call "polygamodioecious." Indulging in a sort of botanical polygamy, some trees bear perfect flowers possessing both male and female parts. Most black gums, however, are dioecious, with individuals producing only male or female blossoms. It took a while until I learned these facts, and it took still longer until I saw not just our black gum's flowers but the configuration indicating their gender. The greenish flowers, densely clustered but so tiny they're well-nigh invisible, bloom on long stalks here and there amid the dark, glossy leaves. Female flowers are more sparse and slightly, very slightly, larger.

And there was another problem to be overcome before I could recognize the black gum in our yard for what it is—a problem of perception, a problem that rose from paying attention to popular wisdom rather than looking directly at the tree: that particular black gum seems to controvert a Freudian principle. Rounded, wider than it is tall, it looks like a breast, not a phallus. That's not the case with the slender, sky-reaching female black gum in which the birds assemble peaceably every September. But bird tree, yard tree—the important thing is that they're both *trees*. As such, they're exempt from man-made principles.

I hear the argument: dryads are an idea conceived by humankind. Note, though, that they are not an idea intended to explain the vagaries of the human condition. And come to think of it, the gender of dryads doesn't matter; be they male, female, or both in one, corresponding with the trees to which they are attached, their true significance lies elsewhere.

At any time in any place since the greening of the world, where trees rise, there is life. And dryads, whatever else they may do, personify that life. Consider them, then, as a means to understanding a rooted way of being that's not remotely like the ambling, scrambling, hither-and-thithering of humankind. Dryads put treeness—that huge, alien vegetable vitality—into terms that mere mortals can apprehend. A friend, who makes her living as a botanist, has mourned the recent felling of a black gum four feet in diameter—that's more than twelve feet around—by a road-construction crew: "If trees had big eyes like whales, dolphins, deer, owls, sea turtles, and feral horses, perhaps people would be more interested in protecting them." But dryads do have the power to give the trees big eyes, and give them ears, hands, feet, rustling voices, and beating hearts. And another greater gift lies beyond these: if each dryad acts as a link between heaven and earth, then each tree draws divinity into itself and makes the sacred manifest.

Dryads in the New World, oh yes, as many as there are living trees. And our yard tree is not just alive, he's flourishing. That black gum stands to last longer than I shall. It's not my place to deny him a dryad of his very own.

THE WOMAN BECOMES A DRYAD

Twelve times since she moved
into the frame house, earth
has reeled around the sun:
twelve falls, twelve burgeonings,
while staid backyard trees
walked closer to her door.

Next spring the doorposts
will bud and sprout,
next spring her spine
shall burst into leaf.
Already her feet are dug in,
her toes put down roots.

THE TREE THAT CAME OUT OF THE SHADE

Loblolly Pine

■

Mud-puddle tree, tree for birds and bull-hunchers, tree that pays the rent today and will figure tomorrow as a major component of the South's fourth forest—loblolly pine is all of these and more. In the experience of some people, it's a tree to sneeze at. Others think that it's nothing more nor less than a pernicious weed. It has served also as a reason for running people out of town.

That's what happened to Bill Stanley in the spring of 1924, when the sheriff hustled him right abruptly out of Conway, South Carolina. Conway was, and still is, the seat of Horry County, a coastal enclave now famed and much visited for seaside vacation resorts like Myrtle Beach. But back then it was a hinterland, a deep backwoods, a wild place where even the bears might get lost and anything could happen. Rednecks and renegades naturally hunkered down in the clearings and supported themselves by working at the things that people have always done in the piney woods and cypress swamps: hunting, fishing, logging, and sawyering, not to mention running off a mort of moonshine. And what was a nineteen-year-old North Carolinian doing down there in rough, tough Conway? Most likely, Bill would have described his job as "hiring." The mildest of the sheriff's words might have been "stealing." I have visions of press gangs and shanghaied sailors, although the actual situation was not quite so dramatic. Bill was simply trying to entice hands away from the Horry County sawmills by promising them better lives if only they'd put their energies to work for the Waccamaw Lumber Company of Bolton, a rural crossroads community—sawmill, plus farms raising corn and tobacco—in Columbus County, North Carolina. The need to do such recruiting says much about the scarcity of mill hands and the hardships, the dangers that were their lot in life. But unceremonious expulsion from Conway hardly discouraged Bill's efforts. He simply traveled on to Marion, South Carolina, where he stopped long enough to recruit some mill hands and also write a letter of yearning apology to his eighteen-year-old sweetheart: "My dearest Lillian, Please forgive me for not coming to see you last week. I had a little trouble with the sheriff in Conway." The problem was described; the letter concluded with protestations of never-failing love. (They would marry a year later and in short order produce five children, of whom my husband, the Chief, is the last born.) It must be recognized, however, that attempts at hiring were merely the proximate cause of Bill's being run out of town. The real reason for the trouble—its bole, its heart, its very root—was the timber that brought the mills into existence: loblolly pine.

Loblolly—an odd sort of name. It would seem to fit the roly-poly rocking toy you can't knock over far better than it fits a tree. But the name makes a certain sense if you can imagine this scenario from North America's colonial days: English settlers, newly arrived on the swampy, densely wooded southern coast, have put their backs and energies into the hard work of clearing farms and building towns amid great, sky-high stands of pines that rise right out of the marshes and mud. The mud is as thick as the porridge—the "lolly"—that sustained them on their voyage across the Atlantic, and like porridge boiling in a pot, it "lobs"—spits, slurps, bubbles, and sucks—as the settlers move slowly through it, pulling first one foot, then the other, from its gluey grip.

Though the true case may be slightly otherwise, it's certain that the colonists noticed from the beginning that this particular pine would grow quite happily in a great loblolly stew of muck and mire. And loblolly has been the most common of this pine tree's common names for more than two centuries. It was given official recognition no later than 1760, when legislation passed by the General Assembly of the Georgia Colony referred to "Squared Timber that shall be made of swamp or loblolly pine." The species, however, is not so fussy that it insists on moist places for its thriving. Though wet conditions make for optimum girth and height, loblolly proves the adage that trees grow not where it's best, but anywhere they can. These pines shoot up almost as vigorously out of well-drained uplands, dry sandhills, and rocky outcrops as they do from floodplain puddles, and they cover their native range in the South like a homespun blanket—thin here and thick there—that extends from southern New Jersey along the Chesapeake Bay down into Florida and westward across the Gulf states into droughty East Texas.

Loblolly isn't the only name by which the tree has been known. Along the centuries, English-speaking people have called it New Jersey fir, North Carolina pine, bull or rosemary or oldfield pine—the last three giving nods respectively to the tree's size, the resinous fragrance of its needles, and its upstart habit of taking over abandoned fields. A bad habit, some would say. Loblolly was noted for its weedy behavior back in the eighteenth century by John Mitchell, the physician, friend of Linnaeus, and medicinal botanist for whom *Mitchella repens*, the partridgeberry, was named; scorning loblolly as "the most pernicious of all weeds," he backed up his opinion with the observation that "they have a wing to their seed, which disperses it everywhere with the winds, like thistles, and in two or three years forms a pine thicket, which nothing can pass through or live in." And to this day, Dr. Mitchell's dismissive view is echoed in his very words—"pernicious weed"—by some otherwise respectable people who give their fealty to other kinds of trees. New Jersey fir tree, a vernacular name now long out of fashion, was placed on the loblolly back in 1748 by Peter Kalm, a Swedish botanist who has left a detailed, eminently readable record of his travels through New England and along the seaboard into Virginia as he searched for hardy plants that would grow well in the short, cool summers and long, cold winters of his homeland. Needless to say, he did not take the heat-loving New Jersey fir back to Sweden. The name "North Carolina pine" attests to the loblolly's economic importance in the Tarheel State. And scientists know the tree by the binomial that Linnaeus himself assigned: *Pinus taeda*, which (Linnaeus notwithstanding) is not quite accurate. It translates as "pitch pine," but though the loblolly can be coaxed into yielding pitch, tar, turpentine, and rosin, it comes in a poor and reluctant second to the southern longleaf and slash pines as a bountiful source of these products, which are also referred to collectively as naval stores for their once significant role in keeping wooden ships watertight and afloat. And places like L. L. Bean that cater to lovers of the great outdoors never use loblolly for the resin-rich, incendiary lightard or fatwood that they sell in sacked or crated bundles for more than a dollar a pound; the kindling that these businesses send out to their customers is taken from the stumps left after the felling of longleaf pines.

For this country's first three centuries, the longleaf pine put loblolly in the shade, causing people to see the latter as a lesser conifer. To this day, longleaf—formally known as *P. palustris,* which means "swamp pine"—is a species that elicits not just admiration but outright love. And in the opinion of its most biased partisans, it is a tree far better than loblolly not just for the excellent lumber it provides and the profit from the naval stores it produces in abundance but also for its towering attractiveness. Looking, listening for whatever comes along, I've spent hours in longleaf glades here and there on the Carolina coast. The typical setting, more properly called a savanna, is characterized by an open understory in which wiregrass grows, in association sometimes with cane and loblolly bay. My progress there is not impeded by the kinds of tangles and thickets found in loblolly stands—the honeysuckle, greenbriar, and poison ivy; the sumac and devil's walking stick. Longleaf depends on fires, once mostly natural, now managed, for its pure stands in a parklike habitat. I walk and look: the pliant needles of longleaf reach from my wrist to my elbow; the plump twelve-inch cones are longer than my booted foot. When the tree is in its infancy, or "grass stage," with no visible trunk, it looks like a giant green shaving brush made of slender needles held close together at the base but spreading full and lush at the top, and so it remains for several years while it puts down a strong, firm root. When the tree is mature, trunk ramrodstraight and the nearest branches unreachably high overhead, it may attain a noble height of one hundred feet. And as it grows old, it may be attacked by red heart, a fungus that causes decay in the tree's heartwood. For all that the fungus may sound like a bad thing, it issues a life-giving invitation to the now rare red-cockaded woodpeckers, which drill first through living wood, then excavate the rotted interior for their nests. The woodpeckers' fate is linked to that of old, fungus-infested pines, not just longleaf but also loblolly and others—as long as they grow in the open habitat maintained by frequent fires or in poor, sandy soil with little underbrush; the fewer the trees that have lived long enough to get red heart, the fewer the nest sites for a small but choosy bird.

Longleaf pine once served as a major focus for commercial lust. For several hundred years, from the earliest colonial days well into the early 1900s, it was a species of grand importance to the American South's first forest. ("First forest" and its numbered successors are terms coined in the late 1960s by a forestry trade group and given an imprimatur in 1988 by the U.S. Department of Agriculture. They are general terms, yes, but serve as a handy means of organizing information. They also work nicely to lend intelligible order to some complex, overlapping, inherently untidy natural processes and to human attempts at their management.) The first forest comprised the seemingly endless virgin stands of longleaf, loblolly, and other pines; of baldcypress and Atlantic white cedar; of hardwoods like oak, maple, and yellow poplar. The second forest was constituted of second-growth timber that had regenerated slowly and naturally while the first forest went under the ax. Now largely gone, it has been succeeded by a third forest, which consists of both regenerated woodlands and plantation-grown timber. The fourth forest only now begins to rise; I'll describe it shortly.

The first forest excited a frenzy of activity from the moment that Europeans first laid eyes on the great boles, the immense crowns, and calculated the yield in lumber or resin products. Centuries of commotion followed the *skreak* of saws, the *thunk* of axes, trees cracking and crashing to the ground, and always men turning the air blue with a great grunting, shouting, cursing racket as they wrestled felled timber out of the woods and the swamps. Men could stand knee-deep in the swamps all day to cut that timber. And when it was down, they'd hitch up oxen to the huge trunks to skid them out of the muck and into the rivers, where the timber was rafted and eventually moved by water to the sawmills. (Listen to those oxen snort and bawl.) But "skid" is far too slick a word to describe what really happened—the humping, tugging, dragging, pushing, shoving that, for the oxen and the effort involved, some loggers called "bull-hunching," a method that lasted into the early 1880s, when patented machines took on the chore. In the earliest colonial days, lumber was manufactured in two-man saw pits, which were simply holes dug in the earth; the log was placed in a rack, with one man standing in the pit and the other stationed above to work the saw. Then water-powered sawmills came along, their machinery driven by the rush of inland rivers or the surge of ocean tides; they, in turn, were superseded by steam-powered mills, and tramways were built to take the timber out of the woods. (The small steam engines chuff and clank. The ripsaws, gangsaws, bandsaws scream as they tear into raw wood.) The sturdy lumber from longleaf pines saw a multitude of uses, from masts and spars to bridge construction to railroad ties and on to flooring. As for the manufacture of naval stores, it was the longleaf and slash pines that were tapped live for the resin that would be distilled to produce turpentine and rosin. (Often as not, however, the repeated tapping killed them.) Tar was made by slowly burning the heartwood and branches of felled trees; pitch, a more concentrated form of tar, came from boiling the black, sticky stuff in a kettle to evaporate its more volatile elements.

The heyday of tapping, logging, and sawmilling the South's first forest lasted for six long decades, from 1870 to 1930 (only a few short years after Bill Stanley was run out of town). Fortunes were made. In 1907, one enterprise operating in Virginia and eastern North Carolina, the John L. Roper Lumber Company, calculated that its timber acreage covered 1,250 square miles and held four billion board feet of longleaf, cypress, Atlantic white cedar, black tupelo, yellow poplar, and a species that the company called "shortleaf North Carolina pine." An article in the April 27, 1907, issue of *American Lumberman*, a trade newspaper, described the Roper Company's holdings this way:

> The casual reader can scarcely take in at once figures of such magnitude, so that more fully to realize the quantities of timber owned by the company it might be said that if its timber were manufactured into boards and plank it would be sufficient to build around the world at the equator a fence of solid boards one inch thick and ten feet high and have enough left to plank a board walk 100 feet wide and two inches thick across the United States from the Atlantic to the Pacific ocean.

This paragraph goes on to cite other boggling statistics, such as the fact that 1,250 square miles encompass more territory than the state of Rhode Island. It also observes, in a typical turn-of-the-century fashion, that the company's holdings "would be represented . . . by the quantity of timber required to make coffins and cases for burying each man, woman and child in the United States at the present time, allowing fifty feet of lumber to each." And it ends with the cheery thought that, because the company intends to maintain its vast holdings, no one need fear exiting this world without a coffin.

The tree called "shortleaf North Carolina pine" was, of course, none other than the loblolly. By 1907, this pine that had formerly lagged far behind longleaf in the esteem of lumbermen was, according to the report, "no longer an article requiring explanation or apology." Indeed, it was the tree that put the butter on the Roper Company's bread, accounting all by itself for a whopping eighty percent of the company's production. The wood, much varied in grain and color, was put to a multitude of uses, from interior paneling and flooring to the manufacture of crates and boxes. The sawmills rose apace, as did the planing mills, dry kilns, and storage sheds. More logging railways were built to bring timber from the woods, and more barges to push mile-long rafts to the riverside mills. And throughout the South companies like Roper added crashing trees and hissing steam engines and screaming saws to the din. Among the larger were the Richmond Cedar Works, headquartered in Richmond, Virginia; the Burton Lumber Company of Charleston, South Carolina; and the Camp Manufacturing Company, with interests from the Dismal Swamp through Wilmington, North Carolina, and on into Florida. The last is an antecedent of the Union Camp Company.

The disappearance of the South's first forest took three hundred years. It happened in slow increments, however, and the new trees of the second forest readily planted themselves where old trees had been cleared out or hardscrabble farms had been abandoned.

The second forest included loblollies, of course, and a few prime specimens stand at Great Neck Point, my home on the banks of the wide and salty river Neuse. They must have been small trees not yet worth cutting when the Roper Company ran its Winthrop Mill over by Adams Creek, a short four-mile bike ride from where I live. The Winthrop enterprise was far more, however, than just a logging operation, along with the obligatory tramway, sawmill, loading shed, and waterside shipping shed; it was a town, complete with company store, company doctor, and company-owned housing for its hands. Today there's nothing left of this sizeable enterprise, except for a few older loblollies, giants among the newer growth, and a relatively open swath cutting through the woods that still cover much of the Great Neck peninsula. The swath was cleared for a tramway that took big timber from the heart of the old forest straight to the mill. No rails are left now, and the woodland surrounding the empty track is young, composed of hardwoods and loblollies, most of them striplings, that have grown up higgledy-piggledy in the decades since the Second World War.

One entrance to the track hides behind roadside weeds only two miles from home. I park my bike on the road's grassy verge, cross the deep ditch with great care (I'm good at falling into such things), part the weeds with equal care (greenbriar has thorns), and walk the wide avenue. It's like the nave of a cathedral, columns of living trees on either side, their branches arching overhead. In fall and winter, woodcocks fly up; in spring, prothonotary warblers flash gold amid the green leaves and sing their loud, buzzing, one-pitch love songs. In summer I avoid the place: beavers have recently dammed a small stream, and its overflow, swamping the path for a good hundred yards, makes a dandy hatchery for the coast's fierce and piratical mosquitoes.

The survivors of the second forest grow closer to home. One grand loblolly stands in the woods, amid cypress, live oak, and fan-leafed dwarf palmetto, only four hundred feet upriver from our land. Some of its branches wear ragtag beards of Spanish moss, blown in by the wind. Several other really big loblollies rise, amid sweet gum and myrtle, in the southeast corner of our own yard—the spot that we watch yearly for the appearance of golden chanterelles. In a fit of precision, I take a steel tape to the tree in the woods and the largest of those in our yard. It's not easy to get the measurements—both trees are corded with Virginia creeper and grapevines as stout as the hempen lines on the old-time wooden ships that, in the days before Roper and Winthrop, used to take trees like these by water to the mills in New Bern, twenty-odd miles upriver. Walking slowly around each tree, steering as clear as possible of greenbriar and poison ivy, I thread the tape between the vines and the red-brown plates of bark. The yard tree has a hearty circumference of six feet, vines not included. The woods tree comes in at eighty-two inches, or six feet ten!

According to the 1994 *National Register of Big Trees*, compiled by the American Foresty Association, the biggest loblolly in the United States is an Arkansas tree, 148 feet tall with a crown spread of 83 feet and a circumference that measures out at 188 inches—15 feet 8 inches. In a somewhat arcane calculation based on these vital statistics—circumference in inches plus height in feet plus a fourth of the average crown spread in feet—this, the country's biggest, specimen of *P. taeda* musters 357 points overall. Virginia's top tree falls somewhat behind that total, but with 177 inches of girth, 135 feet of height, an 80-foot crown spread, and 332 points, it too is far bigger than the trees at the Point. Its massive trunk, however, is oddly contorted, looking as if a mighty hand had grabbed it and given a hard squeeze, with flat, bark-covered shelves of wood extruded between the fingers like dough. But if Arkansas's tree scores a perfect ten (at least for 1994), North Carolina's prize loblolly rates 9.999. Located in Bertie County on the northeastern coast, it boasts a 193-inch circumference (more than 16 feet!), a height of 144 feet, and a crown spread of 77 feet, for a score of 356 points, only one point fewer than those of the national champ.

So, my own big loblolly pines turn out to be pikers. But they're grand, nonetheless, in comparison to most of the local *P. taeda*. And the species grows everywhere at Great Neck Point—in the woods and hedgerows, at the edges of the creekside marshes, in the overgrown field on our back line, around the drainage pond that a neighbor installed for his heat pump. And they're a major, if self-appointed, element in the landscaping of almost everybody's yard.

Not just at the Point but throughout the entire South, loblolly is a persistent—some might even say pushy—fact of life. When the tree sprouts, it looks like a set of long, delicate green whiskers. From that moment on, almost no time seems to elapse before the species, one of the fastest growing pines, develops trunk and limbs and thrusts toward the sky. Like all pines, loblolly prunes itself as it gains in height, sloughing its lower branches as the newer upper branches overshade them. The sticks and stumps of dead branches attract borers and other insects, which in turn summon brown-headed nuthatches, flickers, and the downy, hairy, red-bellied, and pileated woodpeckers; spring, summer, and fall, our yard resounds with tapping, knocking, hammering as the birds come to feast in the bountiful loblollies.

Hardwoods clone themselves easily from suckers. Loblollies and other pines don't. Therefore, to perpetuate their kind, they must rely almost solely on sexual reproduction. Each loblolly has both male and female parts, pollen producers and seed makers, with the former situated low on the tree and the latter, high aloft. This separation of the sexes, ground floor and balcony, is the strategy seized upon by pines to avoid self-pollination and thus ensure genetic diversity within the species. It is the wind, rather than an insect or a bird, that acts as the agent for pollination. The loblollies bloom toward the end of March, when the yellow-throated warblers arrive from the tropics and begin to stake out their nesting territories with *sweet-sweet-I'm-so-sweet* song. As the male flowers shed their pollen, the spring wind goes to work, flinging the stuff through the air like an endless length of pale yellow gauze. Pollen settles on roofs and porches; it covers cars; it floats on the river and washes up on the shore, a muddy tide wrack plastered over sand, shells, and pinestraw. A goodly number of the Point's people are afflicted with clogged sinuses and watering eyes; for a full month, the sound of sneezing is heard through the land. To my good fortune, I'm not among those who suffer but go freely into our pollen-clouded yard to fill paper grocery bags with cones that our dozen-plus loblollies shed in great abundance during the winter; they make fine tinder for starting summer fires in the outdoor grill.

The bigger loblollies in our yard are members in good standing of the second forest. But, like many people on the coast and inland, too, we at the Point are surrounded by acre on acre of third forest. Third forest rises like a sturdy wall beside the roads; it serves as dark green backdrops to the corn and soy fields. Sometimes whole sections of it disappear in a weeklong clanking and roaring of big machines that leave only stubble in their wake.

The third forest may be viewed in two ways, according to Joe Hughes, an environmental forester for the Weyerhaeuser Company, which has established a sizeable presence on and near the Point, with tree plantations covering thousands of acres and a pulp mill in operation twenty-odd miles up the Neuse. Most simply, the third forest is a young, transitional forest. In the broader sense, it may be defined as the frequently unthrifty natural regeneration that follows the logging, mainly on private land, of the second forest's excellent hardwoods and conifers. More specifically, however, Joe describes the third forest as "the pine plantation resource, the so-called 'man-made' forests that are now emerging as a wood supply." From a modest beginning back in the 1920s, such planting, mostly on the coastal plain, expanded by 1990 to a huge fifteen percent of the South's total forest area and an equally boggling forty-one percent of the pine area.

No need to guess which pine is most planted. Joe writes, "Loblolly pine, more than any other species, became the second forest, is the third forest, and is the strong favorite for the fourth." Loblolly now accounts for a monumental sixty percent of all the bareroot—that is, nursery-raised—seedlings planted in the United States. In the South, it is number one, at ninety percent of all pines planted. In North Carolina alone, the species far outstrips all other trees, not just pines but hardwoods, too. The numbers are as big and boggling as those found in the federal budget: 1.5 billion live loblollies, representing 6.6 billion cubic feet of standing wood and 23.9 billion board feet in trees of sawtimber size. Clearly, loblolly puts a lot of rent money not just in the pockets of the timber companies but into the hands of foresters and loggers, truck drivers, sawyers, engineers, pulp mill workers, lumber dealers, and everyone else who makes a living in a trade connected with trees.

Loblolly, the tree that came out of the shade—the reasons for its emergence, its current dominance, are many. It's genetically diverse and resists disease better than many other pines. It thrives on almost every terrain throughout its native South. It's relatively inexpensive to plant. It grows with astounding speed to harvestable size—25 years for pulpwood, 30 to 40 years for sawtimber.

The habit of rapid growth is the force behind the disappearance of the third forest's loblolly plantations. The moment the trees are ready, the timber companies come in, first to thin a stand by removing some trees and creating a series of straight, sun-filled alleys, then to fell the rest after they've put on a bit more height and girth. Afterward, replanting often occurs; the slash in the clear-cuts is piled to disintegrate naturally, and bare-root loblollies are placed in the ground in long, neat, green-whiskery rows. Hawks hunt over the cleared spaces; bears nap in the thickets at clear-cut's edge. Shining sumac shoots up, along with cottonbush and myrtle, and wildflowers like blazing star. But in five or six years, the young loblollies have stretched out their branches and lifted their brushy tips well above the other vegetation. On a small scale, they offer a sturdy green sample of what the fourth forest might be like.

Joe Hughes calls the South's fourth forest "the wood supply of the future." And even more important—more exciting—he expresses the educated hope that the fourth forest will help to provide "a significant solution to the atmospheric problem of excess CO_2" and other greenhouse gases. The reason is one we all learn in school: trees store carbon. Not only that, but they do an exceedingly good job of it; for each pound of tree growth, half a pound of carbon is sopped up from the atmosphere. While the tree lives, carbon is locked up. It also stays imprisoned in processed lumber and is released only through decay. The solution, however, won't be arrived at easily. The American South's fourth forest cannot rescue the world by itself but should be part of a global effort.

At home, however, a thicket of problems grows along with the loblollies, and just as vigorously. Prime among them is the fact that large-scale pine plantations reduce biodiversity. This monoculture—cornfield forestry as it's sometimes called—centers on a single species and so decreases the area's ability to cater to the different habits and tastes of a number of plants and animals. Another troubling question, focused on the loblolly's preferred habitat, is that of wetlands management, of defining wetlands in the first place. And other thorny issues arise, like assessing the effects of intensive herbicide use in the pine plantations and preserving living space for the red-cockaded woodpecker and other imperiled species. Always, human interests need to be somehow balanced with the often competing, make-or-break requirements of the natural world. Now and tomorrow, these matters must be carefully studied and addressed. Nonetheless, the fourth forest, tomorrow's forest, begins to grow beyond mere planning. And the tree that will be tomorrow's mainstay is none other, of course, than the loblolly pine.

At this moment, however, I look to the past. Bill Stanley must have been right downcast and disappointed on his involuntary departure from Conway—no hands for the sawmill in Bolton, no date with his sweetheart. I imagine he was also pretty riled at the sheriff but too mannerly to sass the law—and take the risk of spending time in jail. From my point of view, though, he wasn't run out of town with no reward for his trouble. He couldn't have known it, but he started something back then. This story is for him.

SHAKE THEM 'SIMMONS DOWN

Persimmon

■

For ten months of each year, from winter well into autumn, the common persimmon is a modest tree. Here in my home territory on the lower Neuse, it grows abundantly but unobtrusively, sequestering itself at woods' edge or in the hedgerows amid loblollies and sweet gums, wax myrtle, and sumac. Oh, it does command brief attention in spring, in small part because of the delicate come-hither fragrance of creamy white blossoms. Much more noticeable, though, is the sheer loudness of a persimmon when it blooms. For a solid week, the tree hums, drones, and buzzes. The reason for the din is bees, of course-hordes of bees visiting the blossoms to gather nectar and go about their unwitting business of pollination. But when this short season has faded, the flowers gone and all the noisy nuptial celebrants flown away, the tree again retires from public notice.

Till October, that is. Then, ripening fruit begins to blaze in the hedgerows; amid the glossy dark-green leaves, the fat red-orange berries (as a proper botanist would call them) smolder and flare like live coals. And into November, well after the last leaf has fallen to the ground, those berries still burn on the branches, glowing now with the darker warmth of ash-dusted embers.

What happens next is told in a good, old-timey southern song, a bouncy reel tune better known for its chorus of "Bile them cabbage down" than for its many verses chronicling the adventures of creatures like raccoon and possum, jaybird, and frog.

Possum up a 'simmon tree,
Raccoon on the ground,
Raccoon say to the possum,
"Won't you shake them 'simmons down?"

But I usually can't wait past Hallowe'en to put myself in possum's role and give the local 'simmon trees a mighty shake. Nor is it necessary to wait for a hard frost before the berries ripen. Hereabouts, they may be ready from the onset of October through December. Watch for softening and changes in color.

The objective, of course, is good eating. Fresh persimmons, persimmons baked into bread or cookies—sumptuous treats! And when I gather them, I'm simply one of the latest in a long line of gatherers who have eagerly harvested persimmons along the banks of the lower Neuse for something like the last eight thousand years. The prehistoric inhabitants of this rivershore, people who were forerunners of the tribes later known as Neusiok, Coree, and Pamlico, must have savored that special wild sweetness. It's a safe bet that, October through January, they shook down those sugar-packed 'simmons whenever they had a chance. They surely ate some of their bounty on the spot and may have saved the rest for pulping and mixing with crushed corn to make a kind of bread.

Nor was it especially hard for those early gourmets to locate the succulent berries. The common persimmon, *Diospyros virginiana*, is common indeed. Frequenting the high places and the low, it consorts with pines and palmettos here on the sandy, flat-as-a-flounder coast and, just as easily, with oaks, maples, and rhododendrons at cool Appalachian elevations of up to 3500 feet. The tree may be found throughout North Carolina and, indeed, throughout the entire southeastern quadrant of the United States, with outer limits in central Pennsylvania and central Illinois. It is native to North America, along with a kissing cousin, the Texas persimmon (*D. texana*). That cousin, found in Mexico as well as the Lone Star State, bears small, black, edible berries, which may also be used as a source of natural dye. (But beware the black persimmon: it stains face, tongue, and teeth as readily as it colors fabric.) Both Texas and common persimmons generally come in sexually differentiated versions; they are, as the botanists would say, dioecious, with clusters of small male blossoms found on one tree while the larger, solitary female blossoms occur on another. (And the traffic of airborne bees is concomitantly loud and heavy as they buzz between one and the other.)

The world's persimmons—there are several hundred species—all belong to the Ebenaceae, the Ebony family, and their genus, *Diospyros*, is not only large but far-flung, growing for the most part in the tropics or in temperate zones with long, hot, sultry summers. Woodworkers have long prized the family for its close-grained, night-dark heartwood. But though the persimmons are as strong and black-hearted as any other Ebony, their frequently small girth exempts them from being turned into golf-club heads or parquet floors.

In the eating department, the oriental persimmon, *D. kaki*, may be one of the best known of the bunch. Native to central and north China, it has journeyed eastward across the Pacific with great success. Commodore Matthew Perry, who opened trade with Japan, brought the first seeds to the United States in 1856. Nowadays, varieties of the *kaki* persimmon are not only grown commercially on the west coast of this country but are abundantly available to backyard orchardists through local nurseries and garden catalogues.

It seems not quite reasonable to apply the word "berry" to a fruit as voluptuous as that of the *kaki* tree—fruit to which the hybridizers have now given a wonderment of shapes and colors, from oblate spheres of rosy gold to orange globes containing chocolate flesh and on to plump tomato-red cones. Whatever the variety, all *kaki* persimmons have the smooth skin and palm-filling heft of a nectarine. All make the common persimmon, which is no bigger than a cherry tomato, look like a piker. And, quite apart from their handsome size, some Oriental varieties—not all, but a goodly number—possess a significant advantage over their puny American relatives: a total, blissful absence of the pucker factor.

That infamous astringency is caused by tannin, the substance that makes acorns bitter. And it was remarked in the early seventeenth century by Captain John Smith, who provided the first description in English of the common persimmon—indeed, of any species of persimmon. In a brief but fully packed section of his

Generall Historie of Virginia, New-England, and the Summer Isles, published in 1624, he listed "the things which are naturally in Virginia"—the flora, the fauna, and the uses to which they were put by the native Powhatan Indians. And he adopted the word used by this Algonquian-speaking tribe as the common name for a tree then quite unknown in England and the rest of Europe. (The "Medler" —medlar—with which he compares it bears an edible fruit that resembles a crab apple.)

> Plums there are of three sorts. The red and white are like our hedge plums, but the other which they call *Putchamins,* grow as high as a *Palmeta:* the fruit is like a Medler; it is first green, then yellow, and red when it is ripe; if it be not ripe, it will draw a mans mouth awry, with much torment, but when it is ripe, it is as delicious as an Apricot.

In the hope that this New World marvel might be naturalized, Captain Smith took "Putchamins" home to England. The name for the species, *virginiana,* commemorates the origin of these first specimens to reach the British Isles. But in that cool and foggy climate, though the tree itself will leaf out and flower, the fruit more often than not fails to ripen and stays most stubbornly in the stage of sheer torment.

Less than a century later, English-speaking tongues had turned the word *Putchamin* into Persimmon. The modern name was used for the tree and its berry by John Lawson, Surveyor General to the Lords Proprietors of the Carolinas, in his 1709 book, *A New Voyage to Carolina.* In a chapter detailing the natural history of North Carolina's "Summer-Country," Lawson not only included the common persimmon among the fruit- and nut-bearing trees on his list of "Vegetables"—his general term for all plants—but, echoing Captain Smith's comparison, also devoted considerable space to outlining the persimmon's versatility. Nor was he reluctant in his admiration, finding virtue even in the pucker factor.

> *Persimmon* is a Tree, that agrees with all Lands and Soils. Their Fruit, when ripe, is nearest our Medlar; if eaten before, draws your Mouth up like a Purse, being the greatest Astringent I ever met withal, therefore very useful in some Cases. The Fruit, if ripe, will presently cleanse a foul Wound, but causes Pain. The Fruit is rotten, when ripe, and commonly contains four flat Kernels, called Stones, which is the Seed . . . .

According to Lawson, persimmons were also good for more than pursing mouths and cleansing wounds. He observed and described an Indian game "with the Kernels . . . , which are in effect that same as our Dice, because Winning or Losing depend on which side appear uppermost, and how they happen to fall together." (I can also imagine using persimmon seeds, which frequently number more than four per berry, for a game of Tiddlywinks.)

From sweet fruit to ebony heartwood and on to seeds used in play, it seems that every part of the plant except for the skin may be put to some human use. Even the leaves, picked in summer, may be dried and brewed into a tea with the root-beery flavor of sassafras. But many people cannot bring themselves to think of the fruit as food, much less as a delicacy. The unripe berry's fearful astringency has kept the persimmon stuck firmly in the category of things not to be messed with.

Take heart. There are two ways to beat the pucker factor. In the matter of fresh fruit, the easy way out, of course, is to shun the common persimmon altogether, whether it be wild or cultivated, and to find a backyard orchardist who can supply a nonastringent variety of the Oriental *kaki*. But brave and adventurous souls (who may not have the time to search the hedgerows or the woods) might choose to plant a domesticated variety of *D. virginiana*. Though their unripe berries will cause a sure-fire, full-strength case of torment, these tame trees often have two advantages over the reclusive hedgerow stock. First, they are self-pollinating (though for the greatest set of fruit it helps to plant more than one tree). Second, the fruit grows as plump as an apricot, twice the size of its wild counterpart.

Patience, however, is the key to success in eating any variety of common persimmon in its fresh state. Wait until the berries are thoroughly "rotten." If they offer the slightest firmness when touched, they are not ready. But when the skins seem filled with an almost liquid mush, rejoice and bite in, catching the sugar-sweet juice as it runs down your chin. What, though, if those berries can't be touched, what if they're growing way up there out of reach? No need to bypass pleasure. Obey raccoon's command to possum: give that tree a shake. Immature fruit clings to the branches while fruit that is ready comes tumbling down.

The second way to dodge the pucker factor is to cook the persimmons. Only unripe fruit that's eaten raw can purse a mouth or draw it awry. But even if some immature berries have sneaked into a batch of otherwise "rotten" persimmons pulped for cooking, stove heat will banish every last trace of astringency.

For people who fancy nutrients along with flavor, persimmons rate high as a natural source of iron, potassium, and vitamin C. For me, persimmons define the word "ambrosia." Fresh or cooked, they're a food truly fit for the gods. *Diospyros,* the name of their genus, attests to that. The Greek term may be translated as "Zeus's staff of life." To that I can only add a happy "Amen."

EAT THEM 'SIMMONS UP

When raccoon shinnied down the 'simmon tree and stationed himself immediately below, it's likely that he acted out of self-interest—getting first go at the fruit that possum was shaking to the ground. Likely, too, that enough honey-sweet berries were left for possum to get a fair share. Both critters, though, had to grab while the grabbing was good. People are more fortunate: we can eat our persimmons and have them, too. Here's how.

Persimmon Pulp

Extracting pulp is a quick, almost effortless process.

- Wash the persimmons, be they tame or wild.
- Place a colander containing the fruit over a bowl and mash the fruit with a sturdy spoon. The pulp falls readily into the bowl, while skin and seeds are left behind.
- The pulp may be used immediately, or it may be frozen. Because the short persimmon season does not always coincide with a wish to bake, I usually opt for freezing. Pint-size plastic freezer boxes—two cups each—make ideal containers. The frozen pulp may be safely stored for up to three months.

Persimmon Bread

Received from a friend who fancies wild persimmons and finds them in the loftier altitudes of the Blue Ridge, this recipe makes two loaves of a dark, moist bread that rises very little; an inch of dough may puff to an inch and a half. If you can resist the temptation to eat the finished product all by yourself, it makes an out-of-the-ordinary homemade gift. Because it freezes well, the bread may also be saved for special occasions.

Ingredients

2 cups persimmon pulp
1 3/4 cups flour
1/4 teaspoon baking powder
1 teaspoon baking soda
3/4 teaspoon salt
1 cup sugar
2 eggs, slightly beaten
1/2 cup cooking oil
1/3 cup raisins
1/2 cup chopped walnuts or pecans
1/2 teaspoon cinnamon
1/2 teaspoon nutmeg
2/3 cup water

1. Mix all ingredients together. Divide the resulting dough

evenly between two loaf pans.

2. Bake at 350 degrees for 1 hour, no more. As noted above, the bread will be moist, not cakey, after it is baked.

Yield: 2 loaves

Persimmon Cookies

My neighbor Dorothy, four houses downriver, uses the *kaki*-type of persimmon for these drop cookies, which she prepares according to the following recipe, clipped some years ago from the *Washington Post.* Those of us dedicated to gleaning the local hedgerows find complete satisfaction in using wild fruit.

Ingredients

1 cup persimmon pulp	*2 cups flour*
1 teaspoon baking soda	*1 teaspoon cinnamon*
1 egg	*1/2 teaspoon salt*
1 cup sugar	*1 cup pecans, chopped fine*
1/2 cup butter	*1 cup raisins, chopped fine*

1/4 cup water, to be added only *if dough is stiff*

1. Blend persimmon pulp and baking soda. Beat the egg lightly and add it to this mixture.
2. Cream the sugar and butter.
3. Mix the dry ingredients together. Then combine them with the pulp, baking soda, egg, sugar, and butter.
4. With a teaspoon, drop the dough onto a greased baking sheet. Bake at 350 degrees 12–15 minutes, until light brown.

Yield: 4 dozen cookies

Persimmon Pudding

Pity the possum, pity the 'coon, both limited to the fruit they can shake from the branches. They never had a chance to taste this honey of a dessert. The following instructions appear in *Down East Recipes,* a collection made in 1982 to benefit a sadly no-longer-extant rural hospital. The book features traditional offerings from the North Carolina coast.

INGREDIENTS

2 cups persimmon pulp
3 eggs
1 1/2 cups sugar
1 1/2 cups flour
1 teaspoon baking powder
1 teaspoon baking soda
1/2 teaspoon salt
2 cups whole milk
1/2 cup melted butter
1 cup raisins or nuts
1/2 teaspoon cinnamon
1/2 teaspoon nutmeg
1 teaspoon vanilla

1. Combine all ingredients. Pour into a lightly greased baking dish.
2. Bake uncovered at 325 degrees until firm, about 1 hour.

Serves 6 moderate eaters—or 4 of the sort with a raging sweet tooth.

PERSIMMON–SWEET POTATO PUDDING

Dorothy also contributes this less traditional pudding recipe, the source of which has been lost in time or transit. Safe to say, however, that it's southern—note the sweet potato. It's a remarkable concoction in several respects: for flavor, of course, and for the fact that it calls for no sugar other than that found naturally in its beautifully orange main components.

INGREDIENTS

1 1/2 cups persimmon pulp
2 cups raw sweet potato, shredded
2 cups flour
1 cup milk
2 eggs
1 stick (1/4 pound) butter, melted
1 teaspoon baking powder
1/2 teaspoon cinnamon
1/4 teaspoon allspice
1/4 teaspoon nutmeg
1 teaspoon vanilla

1. Mix together all ingredients. Pour into a lightly greased 9- by 13-inch baking dish.
2. Bake uncovered at 350 degrees for about 50 minutes.

Easily serves a dozen dessert lovers.

IN SEARCH OF HARMONY

Yellow Poplar

■

Yellow poplar, tulip poplar, tulip tree, Rakíock, *Liriodendron tulipifera*—no matter what it's called, I've always seen the tree as lofty. And not just lofty but aloof. Straight trunk, narrow crown, its posture is rigidly erect. It holds its leafy branches disdainfully far above my head. The beauty of the flowers, large greeny-white cups lined with touches of saffron and redgold, can't be denied, but the loveliness seems all on the surface, no substance beneath.

Trees usually evoke my wholehearted interest, and if the species is eccentric or even grotesque, so much the better. But the yellow poplar is an alien breed, or alien to me at any rate. It leaves me cold. I've never been able to understand how—and why—it spoke to other human beings and turned them almost unanimously into raving enthusiasts who do peculiar things like calling the tree "supreme for bigness and harmony." Harmony? That's the opinion of at least one fan, Rutherford Platt, who wrote a chatty and wide-eyed book on American trees back in 1968. Twenty years earlier, Donald Culross Peattie went even further, commending the tree for its "stately beauty" and crowning it "king of the Magnolia family." And he found "something joyous in its springing straightness, in the candle-like blaze of its sunlit flowers, in the fresh green of its leaves, which . . . are forever turning and rustling in the slightest breeze; this gives the tree an air of liveliness lightening its grandeur." Sentiments of this giddy sort are hardly uncommon. Even state legislators have fallen head over heels for the yellow poplar: not one but three states—Indiana, Kentucky, and Tennessee—have declared it their official tree.

These people and their numerous ilk seem to think that admiring the tree is a matter of course, with the implication that if one doesn't, one is grievously flawed. But I've never been able to cozy up to a yellow poplar. I've never heard its particular music. Or not till recently, that is, and the process has been no overnight phenomenon, no quick sloughing of ancient prejudice, but rather an event like the slow flutter and fall of yellow autumn leaves.

Whence the widespread human propensity to cheer for the yellow poplar, to give it a completely uncritical thumbs-up? The only way to learn was to try looking at it through the sensibilities of other people. I sought especially comments from people whose observations might exhibit a shade less bias than those of all-out advocates like Platt and Peattie (and might, not incidentally, help me to maintain my accustomed distance).

The first witness called on was Thomas Harriot, member of Sir Walter Raleigh's 1585 expedition, the one that landed colonists on Roanoke Island. In his *Briefe and True Report of the New Found Land of Virginia,* published five years later, he noted the species as something useful for future settlers—"those which shall plant and inhabit"—to know about. Calling the tree by its native name, Rakíock (and giving it a capital letter according to the practice of his day), he proceeded to describe it mainly as a form of transportation:

> Sweet wood of which the inhabitants that were neere unto us do commonly make their boats or Canoes of the form of trowes [trees]; only with the helpe of fire, harchets of stones, and shels; we have known some so great being made in that sort of one tree that they have carried well x x. men at once, besides much baggage: the timber being great, tal, streight, soft, light, & yet tough enough I thinke (besides other uses) to be fit also for masts of ships.

But concentrating as it does on the tree's usefulness to humankind, this tidbit conveys to me no sense of the tree itself. Instead, it imprints a picture of a twenty-man canoe being paddled full speed ahead across the lapping, slapping waves of Roanoke Sound.

Quite another picture appears in the description given by John

Lawson, Surveyor General and promoter of colonial real estate for the British Lords Proprietors of the Carolinas. In his 1709 compendium, *A New Voyage to Carolina,* which treats of such matters as geography, natural history, and the lives of the native inhabitants, he made several direct observations on the "Tulip-Trees, which are, by the Planters, call'd Poplars," and he delighted, too, in handing on a bit of hearsay. His experience speaks in down-to-earth notes like these: that the trees "grow to a prodigious Bigness"; that the wood makes fine wainscoting, shingles, and planks; that an ointment made from the buds is applied to cure scalds and burns; and that cattle relish eating the buds, which then "give a very odd Taste to the Milk." But it's the hearsay that makes my mind's eye open wide. Immediately after reporting that the species may measure as many as twenty-one feet in circumference, he writes:

> I have been informed of a Tulip-Tree that was ten Foot Diameter; and another, wherein a lusty Man had his Bed and Household Furniture, and liv'd in it, till his Labour got him a more fashionable Mansion. He afterward became a noted Man, in his Country, for Wealth and Conduct.

Bigness like that defines "prodigious." But here, in a variation on not seeing the forest for the trees, I couldn't see the tree for the lusty man, his bed, his table and chair, his pots and pans, and maybe his cat and his dog.

Peter Kalm, member of the Swedish Academy of Sciences and collector of American plants, drew singular attention to the yellow poplar's leaves. In his journal entry for October 19, 1748, a short month after his arrival on the shores of the New World, he had much to say about the species, starting off with comments about its ubiquity and its great size. He noted its service to both Indians and Europeans as a "canoe tree." He added considerably to John Lawson's list of uses for the wood—not just shingles and planks but also bowls, dishes, spoons, and doorposts—and mentioned seeing a fair-sized barn with roof and walls built entirely of the boards split from a single tree. The wood is reportedly easy to work, though it contracts enough to crack in hot weather and swells almost to bursting when the rains come. Nor did Peter Kalm fail to note the tree's medicinal properties: a poultice of crushed leaves plastered on the forehead to banish a headache or on the feet to alleviate gout; a decoction of roots to ease or stave off the fever and ague, as malaria was then known; a dose of pounded bark administered dry to horses to rid them of worms. His description of the flowers is secondhand, for he'd arrived in America well after the tree's spring blooming had come and gone (he would gaze on the flowers to his heart's content the following year and discover that "they have no smell to delight the nose"). His words make it clear that he found something strange in the idea of an enormously tall tree covered for two weeks in May with big, tulip-like blossoms. But though he hadn't yet looked on the tree's distinctive flowers, he certainly did see the leaves. He shook his head, picked up his pen, and wrote:

> The leaves have likewise something peculiar; the English therefore in some places call the tree "old woman's smock," because the leaves resemble one.

And with that, all I could see was a smock, a plain green apron with bib and skirt—and the plump, grey, grandmotherly soul who's wearing it (and stirring up a pie or a syllabub for the lusty man). I still had no sense of the tree and how it fits into the lay of the land, how it might find some harmony with the rest of the world (if not with me).

It seemed then that an artist's vision might reveal at least some of the enchanting connections that had eluded me whether I saw the tree in the woods or in the words of others. Not one but two artists quickly obliged. The first was Mark Catesby, the English naturalist who traveled the Southeast in the early 1700s and not only drew many pictures of the New World's amazing birds and plants but furnished commentary. And it was the leafy, twiggy tip of a yellow poplar branch on which he chose to set his "Baltimore Bird," the

Baltimore oriole. The bird—a male, with black wings and the rest "of a bright color, between red and yellow"—hangs on to a twig. Leaves that are bigger than he is dangle from "footstalks, about a finger in length" and almost hide the basket nest in the upper left corner. Above the bird, a cone-like seedpod thrusts straight up; below him, the single flower has opened so fully that it's well past its prime. Mark Catesby thought the flower more closely resembled fritillaria than tulip. (If he'd had any say in the matter, we might speak today of the fritillaria tree.) He did note a close connection between tree and bird—"It breeds on the branches of tall trees, and usually on the poplar or tulip-tree." With that information, I found a good measure of gladness for the oriole. But the tree, lofty and aloof as ever, maintained a quite stubborn distance.

Catesby, granted, did provide me with some small help. But John James Audubon gave none at all. His painting, begun in 1822 and finished in 1825, also places Baltimore orioles, three of them, in a yellow poplar. The female, orange-gold with black wings trimmed in white, clings to the outside of the finely woven nest that she has just built and furnished with softest linings. Two male birds, feathered in flames and midnight shadows, perch in courting postures—Choose *me!* No, *me!*—on nearby twigs. Behind them, amid densely clustered green leaves, the tulip-like blossoms are shown in the bud and in full bloom. The problem here is that the emphasis falls almost wholly on the birds. The painting's botanical component is needed only as their foil. It's a motionless backdrop, an elegant wallpaper, against which the orioles become more than painted figures. Here, in contrast to the perfect stillness of leaves and flowers, they hop, perch, and flap their wings. I can almost hear their songs. Unlike the greenery behind them, the gorgeous birds are preternaturally alive.

Life: of course, yellow poplars have as much claim to vitality as orioles do. They have as much as I do, and perhaps more, for trees as old as I stand an excellent chance (barring lightning and loggers) of lasting far longer than I can. But yellow poplars are quiet about themselves—and as unromantic, as lacking in charm, as dull and tedious, as, well, wooden as any broomstick. And after looking at the species through the eyes of two artists and listening to the talk on the tongues of a gentleman adventurer, a gossip-mongering promoter, and a traveling scientist—all of whom saw it mainly in the light of its utility to people and a few birds—I did not change my mind. Nor had I learned anything about its true character, its treeness, not to mention its place in any larger harmonic scheme. It was high time to call in someone able to see the tree as a *tree.* It was time to consult a botanist.

If ever anyone has been able to dodge the traps of subjectivity and to see trees in their own image, that man was Charles Sprague Sargent. His masterwork, the comprehensive *Manual of the Trees of North America,* was first published in 1905 and issued in a revised edition in 1922. As I'd hoped, it supplies some noteworthy facts, the first of which is that the yellow poplar isn't a poplar at all. It resembles the true poplars only in appearance—the tall, straight trunk, the narrow crown. Poplars belong to the Willow family, the Salicaceae, while the yellow poplar is one of the Magnoliaceae, the Magnolia family (which, as I've learned elsewhere, is among the most primitive gymnosperms—that is, the trees that bear protected seeds—for the seeds of this ancient family are exposed on the outside of their cones rather than enfolded within them). Formally, the tree is *Liriodendron tulipifera,* which means "tulip-bearing lily tree" (lily and tulip—the flowers resemble both, and Linnaeus, who chose the binomial, clearly opted to cherish both names rather than throw out one). Time was, back in the Cretaceous period, that *Liriodendron* thrived in what is now Europe (*Tyrannosaurus* preyed then on duck-billed dinosaurs, and the first forests of deciduous trees rose up tall and green). But today the genus is confined to one species in North America and one, *L. chinensis,* in central China. The American yellow poplar grows in the cold winters of southeastern Vermont and in the steamy summers of northern Florida; it attains great heights in Appalachian hollows and in the Mississippi valley. It flourishes—particularly where the soil is deep, rich, and moist—in the lowlands and also in the highlands up to an altitude of 5000° (that's right, 5000°; it was Dr. Sargent's practice to use the symbol for feet to indicate inches—"Flowers 1½′–2′ deep, on slender pedicels ¾′–1′ long"—and the symbol for degrees to indicate feet—the yellow poplar "sometimes nearly 200° high, with a straight trunk 8°–10° in diameter, destitute of branches for some 80°–100° from the ground"). All well and good—but his description of *L. tulipifera* also overflows with material like this:

> Flowers . . . inclosed in the bud in a 2-valved stipular membranaceous caducous spathe; sepals spreading or reflexed, ovate-lanceolate . . . ; petals erect, rounded at base, early deciduous; filaments filiform, half as long-as the linear, 2-celled extrorse anthers adnate to the outer face of the connective terminating in a short fleshy point. . . .

Trunk, leaves, fruit, and seeds receive the same treatment; the most minute, the most intimate, details are heaped high and spread thick. For anyone who's not a botanist, reading such material may seem an exercise akin to struggling across a parched wasteland, with not one saving drop of rain in sight. But this, I know, is a necessary dryness. It provides objective descriptions that anyone in the field will understand. It keeps metaphors from sprouting as wildly as kudzu and covering up What Is with a lot of imprecise As Ifs.

But harmony depends on balance, progression, multiple parts. It calls for moisture along with drought. And I wanted rain. I wanted wellsprings and juice. I wanted flowing sap. On a briskly blue November day, with fall colors beginning to fade and great flocks of blackbirds swirling through the air like pepper, I went to visit North America's biggest yellow poplar.

It's big indeed. And by some random act of fortune, it grows in Bedford, Virginia, only a short two-hour drive from my Shenandoah valley home. By coincidence, Bedford seemed a most appropriate location. Thomas Jefferson built Poplar Forest, his beloved home away from home, on a 40,000-acre tract of land in Bedford County that his wife had inherited from her father in 1771; there, as a private person rather than a statesman, he could think, write, entertain family and friends, and work in his gardens. Poplar Forest was named, of course, for the specimens of *Liriodendron* that grew in uncountable numbers in the wild, and Jefferson, a man with a self-confessed passion to plant trees, also chose to include it as an element in the landscaping of his hideaway (which boasted grounds and architecture as elegant as those of Monticello). Nor did his efforts on the tree's behalf stop there. Appreciating it not as a provider of boats, housing, or suitable sites for birds' nests, but in its own right as a tree, he regularly supplied great quantities of its seeds to horticulturally minded friends in France.

The big tree, though it does not occupy a place within Poplar Forest's former domain, grows nonetheless in the same distinguished bailiwick. Before I went to pay a call, I saw its photo and learned its vital statistics: 30-foot 3-inch girth, height of 124 feet, 122-foot crown spread. Its age is unknown. The April 1986 issue of *American Forests,* which also contains that year's National Register of Big Trees, had furnished the picture and measurements, along with a full-page spread of awestruck, almost breathless praise. I was told that the tree was a "royal presence" and instructed in proper behavior: ". . . if reverence is due nobility, everyone visiting this tree should kneel." As if such nobility required protection from hoi polloi, directions to the palace were not clear; the article said only that the tree was tucked away in the woods off Smith Street and that the faint path leading to the throne room was indicated only by "a chained-off drive leading to an abandoned shack and a tumbledown dog pen." The photo shows the giant surrounded by a full green court of understory shrubs and other trees. Beside Gargantua, all of them, though hearty in their own right, look delicate, if not downright peaked. (And so does a pilgrim—the writer of the feature—who sits crosslegged on the ground with her back against the majestically furrowed dark grey trunk.) But if I wanted to see the real thing, not just a photo, I needed more accurate directions. They were quickly supplied by the forester at the Virginia Department of Forestry's Bedford office. He warned me, however, that today things weren't quite as they'd been in 1986.

What greeted my eyes was Poplar Park, established in 1991. The venue for the big tree is no longer sylvan. The faint path is now a suburban road named Grand Arbre Drive (though, goodness knows, the yellow poplar was never French). The abandoned shack and empty dog pen have been replaced by two-family frame houses, dozens of them, painted white, grey, or brick red and one or two stories tall. The park, set at the bottom of a dip amid the houses, occupies enough land for two or three small lots. I put the car in a graveled pull-off. Stone benches have been placed here and there on the grassy lawn for the benefit of gawking pilgrims, who cannot now come nearly so close to the noble presence as recently they did. The tree is guarded by a five-foot chain-link fence that encloses another patch of lawn and bears two signs with white letters on brown: NO TRESPASSING. Noon, and the November breeze held a definite chill. The stone benches would be too cold for comfort. Camera in gloved hand, I walked across the grass to read another, more friendly sign about the founding of the park by a committee of ladies, eager to protect the tree when suburbia began bulldozing the woods. The tree itself was discovered by two Bedford policemen. A second friendly sign, using—hurray!—the illustration from Charles Sprague Sargent's *Manual,* gives information about the tree. It has grown since 1986! It's put on another 9 inches around its middle, 3 feet of crown spread, and 22 whopping feet of height for 146 feet altogether.

It doesn't look that tall. Except for its smocklike leaves, it doesn't even look like a yellow poplar. It's not lofty, with trunk and limbs reaching haughtily toward heaven. Instead, six hefty branches jut from the massive trunk only twelve to fifteen feet above the ground. I was reminded, absurdly, of a giant squid: short, thick grey-black body and long, stout tentacles, which, in this case, thrust powerfully upward into the blue sea of heaven. Dark green algae and moss coat the tree's muscular branches on their upper surfaces; leaves—

golden, dry, rustling in the light but nippy breeze—were still clinging in great plenty to the topmost twigs. Several large branches that once extended outward near the ground have decayed and disappeared; they are commemorated now by gaping holes. And the trunk itself gapes. In the side of the tree, perhaps eight feet up, rot has excavated a huge cavity that looks as if it could indeed accommodate a lusty man and his household goods. I took pictures but could not get close enough to estimate the hole's interior dimensions. Instead, I walked down the driveway of a nearby house to get a half-look at the side of the tree that does not face the park. Roll of film finished, I walked back to the car and on the way picked up four samples of the old woman's smock—brown, gold, green-gold, and a red-orange the color of Virginia clay.

The national champion yellow poplar is, I am quite sure, a curiosity. It's an overweight tree, gluttonous and swollen, that has fed with nonstop gusto on mineral nutrients, water, and sunlight—especially sunlight. For sheer size, it's a source of bogglement and fascination, and if size is any indication of status, then it's not merely noble but downright imperial. But it's nonetheless a tree out of context, caged behind chain link, companion trees and understory gone—though fence and park were clearly the only way to keep it safe, once two-legged creatures with central nervous systems and chain saws invaded its domain. It's also a nonconformist, a dissonant tree, that finds little accord with the species as a whole. Nevertheless, the champ has the grudging, curmudgeonly sort of charm that comes with oddity and the serene flouting of convention. It did not, however, thaw my feelings toward its kind. So, I wished it long life and bade it farewell.

Not long afterward, I mentioned my difficulties with yellow-poplar appreciation to my friend Katie Lyle, who often goes rambling outdoors to gather wild foods and then writes books full of lore and recipes. I said, "I don't like yellow poplars. I can't understand them."

"Oh," she said brightly, "you must not know about the morels."

The morel is *Morchella esculenta,* a mushroom so truly succulent that mentioning it—just thinking of it—turns on my salivary glands. It's large, light brown, and long stemmed, with a cone-shaped cap that's embellished with many shallow pits and ridges. And it's found thrusting up through damp leaf mold in the springtime woods. That's all I knew about morels. I've never gone hunting for them, but my father did, and most successfully. So does Katie, every year, and she's been good enough to share her bounty sautéed and wrapped in filo pastry.

Taking pity on my bafflement, she said, "The way to find morels is, you look under the poplars. They have a symbiotic relationship—so that's the best place." And she gave me the name of her companion in the annual hunt, a biologist who teaches at the Virginia Military Institute.

Burwell Wingfield knows about the ways of fungi in the woods. He knows about the ways of trees and other plants, and the ways of animals, too. He knows about their insides as well as their outsides, and the secrets that they hide within eggs and seeds or conceal down in the rich darkness under the earth. He told me precisely what goes on between the yellow poplars and the morels.

Their association is mycorrhizal—an association, that is, between a fungus (the Greek word is *mykes*) and the roots (*rhizai*) of another plant. And their symbiosis, the terms on which they live together, is known as mutualism, a positive relationship in which each one depends for dear life on the other to provide some necessary benefit. It's not an exclusive relationship, however; morels cooperate not just with the yellow poplar but also with ash and apple and even some grasses, while the tree makes itself available to other kinds of fungi. The essential element, the element without which neither one can thrive, is the partnership.

And just how do the morel and the yellow poplar help each other? How do fungi and trees join forces in order for each to live? Year after year, a quiet, dark, quite peculiar event takes place. Deciduous trees store sugars in their bark to protect them from freezing. But in the spring, when the sap begins to run, the sugars descend into the tree's roots and displace some of the water that was there. More sugars, less water—the new proportions alter the process by which the tree takes up water from the soil. It's the fungus in the soil that helps water move into the roots and from there into the sapwood to the upper parts of the tree. And with the springtime arrival of water in higher regions, buds burst open, leaves begin to show green, and flowers bloom. Then, when the leaves reach a certain size and ground temperatures attain a certain warmth, "something magical happens." That's how Burwell Wingfield puts it, and he adds, "Something that has been repeated since before there was a proliferation of mammals, much less developing mankind. Something that probably young dinosaurs sought out every spring. The dump of thiamine and sugars to the roots triggers the formation of fungus fruiting bodies—the morels."

His annual hunt for *M. esculenta*—"moreling," as Katie calls it—is for him a mystical experience. He is transported beyond the ordinary passage of hours, days, and years into a time "older than recorded rituals, older than recorded history, older than mankind, older than independent continents. One is participating in one of nature's basic cycles and it brings *me* much comfort and pleasure."

Yellow poplar, morel, dinosaur, human being—it is impossible not to respond to their resonance. The thrum of a sympathetic vibration finally sets in. I am content. Meanwhile, the yellow poplar goes its own way, not giving a damn.

TREE-KNOWLEDGE

Seed fingers up, tongues darkly down.
Roots anchor, breaking stone.
Bark shields, heartwood holds tall the
middle core
that rings in ample and thin years.
Sap freezes, flows. Buds grow fat,
explode:
leaves springing lightward breathe and,
burning, fall
and seed thrusts home
knowing, to live, true answers for earth,
light, wind, spanworm and storm.

Eyes, what is your sun?
Foot, where the firm ground?
Skin, bones, sinew, blood and lungs,
knowing without season that you die,
you blow, gnaw, bend, ask your own
questions
and tell, to live, vital lies.

SWEET GUM, WITH OCCASIONAL BIRDS

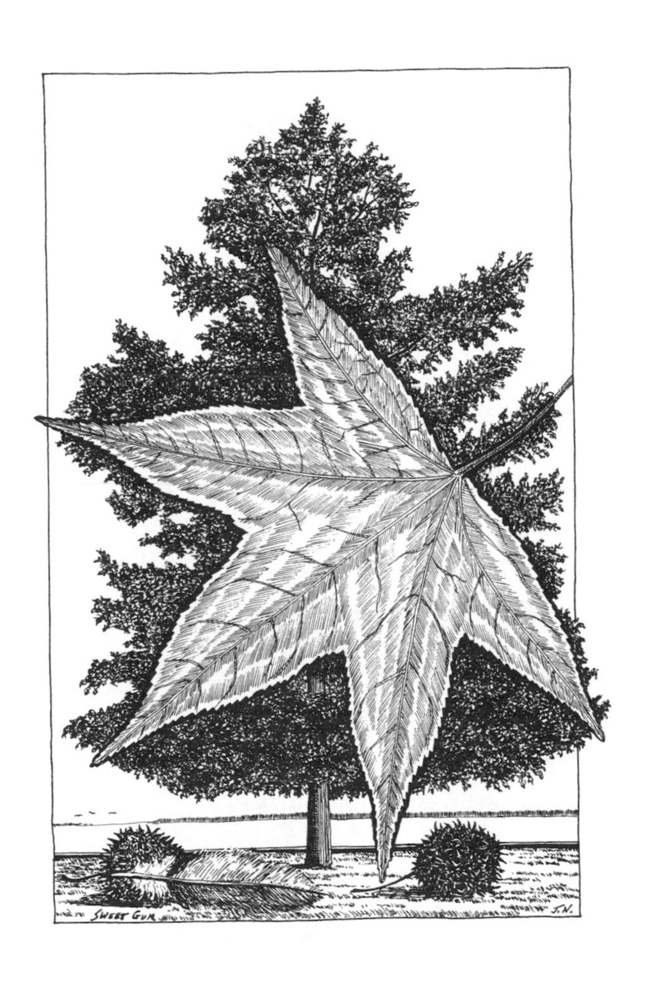

T*ick tick-tick tick tick*—the light but regular sound of tapping comes at fairly close intervals from dawn till dusk: yellow-bellied sapsucker at work. The bird is female; though red feathers crown her head, none show at her throat. The rest of her is black and white, no yellow anywhere, except for the merest hint of ivory on the belly—but only in the right light. Twenty feet off the ground, using her stiff tail as a prop, she inches her way up the trunk of a sweet gum.

To my great good fortune, she has chosen the side of the tree that faces our trailer. She could have selected its river side, for the tree grows just six feet from water's edge. But as is, for three full weeks from the first of March till the equinox, my husband and I can sit at ease on the front deck to watch her as she drills a series of holes in the wood. We also watch the river beyond the tree, and the ducks—mostly ruddies, buffleheads, and black scoters—that are rafted there, staging for migration north. The wintering birds are still with us—ducks, tiny Bonaparte's gulls, yellow-rumped warblers, tail-bobbing phoebes. The kingbirds and peewees, the summer warblers, the yellow-billed cuckoos and chuck-will's-widows that nest here have not yet returned. Right now, the season holds its breath, waiting for change.

The sapsucker is one of the winter species, soon to fly. Though she might have chosen another kind of tree, she's tapping this sweet gum for her journey's necessary fuel, for the sugary juices and the insects caught thereon. How methodical she is, how precise her work! The holes are spaced a quarter-inch apart and arranged in closely set vertical rows. By the time she's done, she's drilled at least a thousand little pits in a section of trunk ten inches wide and three feet high. Liquid seeps out and glistens.

The day before equinox, vast flocks of departing ducks pepper the pale sky. On the day itself, I hear the year's first yellow-throated warbler, see the first gnatcatcher. The sapsucker is gone.

The sweet gum weeps its watery juices for a week, then starts to heal.

II

River and sapsucker—this sweet gum possesses not only physical being but habitat and history. It thrives in its own special context. And its species inhabits a world of time and space as large and colorful as that of any tree, and perhaps more so than some.

Every tree rises, of course, within a world beyond its rooted self. There is a certain region in which it must grow, and a certain climate, a certain soil. It shows preferences in the company it keeps, its own or that of particular other species. It usually belongs to an extended family of near and distant kin, some of which may live in realms as far away as China.

Though rooted, no tree stands still. It stretches skyward and puts on girth. Budding, burgeoning, setting fruit, it expands each spring, sinks back each autumn, sleeps away winter, and starts all over again with the returning light and warmth.

Every tree also lives amid myriad alien lives. Worms and insects putter around the roots. Birds nest in the branches (or drill holes for fuel). Caterpillars eat the leaves. Many creatures make use of the wood—the borers that tunnel in, the woodpeckers that dig them out, owls roosting in holes, loggers cutting trunks for shipment to furniture factories or pulp mills. And the kind of tree each tree is determines the ways in which it fits into its surroundings and serves a host of purposes apart from its own.

And every tree occupies a niche in time; even if no one notices, a history accrues to it like interest on a lifelong loan. A tree's years may be counted not just by growth rings but by events. Some are natural: lightning; fire; fungus; too little rain, too much, or just enough. Deer may scar the bark when they batter at it with itching antlers; bears rub it off when they scratch their backs. (And woodpeckers will drill holes.) As for events in which humankind figures, they range from pruning to pinning on scientific names.

A tree may also stand in a context that no one can see or touch, a context located in the human psyche. Such a context is no less real, however, than the tangible kind. For me, this aspect of the sweet gum is especially potent: The species sustains contemplation in a most satisfactory way. But I may be biased.

III

The past sends bulletins.

1519: The Western world catches its first whiff of the sweet gum tree. The place, Mexico; the occasion, a ceremonial meeting between the Spanish conquistador Hernán Cortés and the Aztec emperor Montezuma II. And when Montezuma smoked a substance enclosed in a highly ornamented pipe, Don Bernal Díaz del Castillo, the expedition's historian, sniffed the air and recorded his impressions: ah, "liquid-amber, mixed with an herb they call tobacco." Though the scent of the latter was new and strange, the spicy aroma of the former must have reminded him of the incense used in churches back home—incense made from the fragrant gum of a Middle Eastern species that Europeans knew as the liquid-amber tree. As shall be seen, Don Bernal's nose has not misled him.

1528: Liquid-amber trees are noted in La Florida, Spanish territory that will, nearly three hundred years later, become part of the United States. They are huge, growing in open woods amid equally great hickories, magnolias, live oaks, cedars, and baldcypress. So says Álvar Núñez Cabeza de Vaca, who will write his account and publish it in 1542, when he goes back to Spain after being shipwrecked and enslaved by Indians, then escaping to slog westward perhaps as far as California before reaching his compatriots in Mexico and finding at last a day of safe return home.

1651: The first detailed description of the New World's liquid-amber tree sees the light of print. The account was written in Spanish some seventy-five years earlier by Francisco Hernandez, an ardent herbalist, who studied the plants of Mexico. Not just the physical characteristics of the tree but its imputed medicinal virtues caught his fancy. He reports that its resin, "added to tobacco, . . . strengthens the head, belly, and heart, induces sleep, and alleviates pains in the head that are caused by colds. Alone, it dissipates humors, relieves pains, and cures eruptions of the skin. . . . It relieves wind in the stomach and dissipates tumors beyond belief."

1709: The sweet gum, by that name, is included on a list of "Vegetables" to be found in "the new-discover'd Summer-Country" on the mid-Atlantic coast. The list maker is John Lawson, who is not only Surveyor General to the British Lords Proprietors of the Carolinas but a salesman making a smooth, if hyperbolic, pitch for New World real estate. (He is also an intrepid adventurer, destined to die two years later at the hands of the Tuscarora Indians.) Clearly, he admires the tree:

> The sweet Gum-Tree, so call'd because of the fragrant Gum it yields in the Springtime, upon Incision of the Bark, or Wood. It cures the Herpes and Inflammations; being apply'd to the Morphew and Tettars. 'Tis an extraordinary Balsam, and of great Value to those who know how to use it. No wood has scarce a better Grain; whereof fine Tables, Drawers, and other Furniture might be made. Some of it is curiously curl'd. It bears a round Bur, with a sort of Prickle, which is the Seed.

Except for the concluding squib about the seed, little weight is placed on the tree's botanical characteristics. Nor is heed paid to beauty—the intricate and lovely architecture of that seed, the leaves like five-pointed stars, the annual miracle that turns their leaves from green to red and midnight purple. Speaking in a voice not untypical of his day (nor of the earlier times of Francisco Hernandez), Lawson zeroes in on the species' pure usefulness to humankind. Its resinlike gum is said to heal skin diseases that scar and harden subcutaneous tissues (the "Morphew") or cause the blisters and itchy scaling of such ailments as ringworm, impetigo, herpes, and eczema (all of which were considered "Tettars"). And the tree's hard, close-grained wood puts joy in the eyes of cabinetmakers and lends eagerness to their shaping hands.

Nowadays, nearly half a millennium since Don Bernal trusted his nose and three centuries since John Lawson delivered hispraise, the gum has long ceased to serve as a tobacco additive or as a remedy for everything from tumors to eczema and flatulence. But the makers of furniture (and barrels and boxes, plywood and fine veneer) still rejoice. And so do all, throughout the tree's ancient domain, who love cool green shade.

IV

Sweet gum would not be sweet nor gummy without its juice. The tree's everyday name and its more formal appellation both acknowledge that aromatic substance. *Liquidambar styraciflua*—that's the binomial chosen by Linnaeus. The genus name marries the Latin *liquidus* with *ambar,* derived from Arabic, and it means just what its look and sound would indicate: "liquid amber," for the appearance of the juice. The English for the species name, which tacks the Latin *-flua* onto a Greek beginning, is "flowing storax."

Storax? Does that mean the sap? Sort of. The word "storax" actually refers to the gum, and specifically to the gum obtained from another member of the genus—*L. orientalis,* "liquid amber of the East," a native of Asia Minor. And where its North American counterpart weeps in slow, reluctant trickles, the oriental sweet gum delivers its juice in commercial quantities. Both perfume (the spicy scent noted by Don Bernal) and expectorants are made from this fragrant, grey-brown balsam.

The gum is one product of sap. Think of a tree as a great green refinery. The nutrient-bearing liquid, the crude sap, that rises upward from the roots is converted by the leaves into a substance known by botanists as elaborated sap, which usually has a downward flow. And from this elaborated sap, a tree may manufacture a number of substances. Among them are not only gums, resins, and oils but also latex and sugar. For humankind, these products are the stuff of medicine, flavorings, rubber, chewing gum, and syrup for drenching waffles and pancakes. For the sapsucker, which drills its neat holes not just in sweet gums but in many kinds of trees, they provide both primary food and a sticky trap for winged protein.

Impelled by my usual curiosity, I go on a hunt for sweet gum's flowing storax, a substance that has served many uses, from make-do chewing gum to an essential ingredient of modern home remedies for coughs and diarrhea. Books give varied advice but indicate that the gum may be collected from "spontaneous exudation" or extracted by tapping the tree's inner bark. The now dry exudations caused by the sapsucker are far above my reach. I try getting at the inner bark of several young trees in a hedgerow, first with a hatchet to strip off a section of outer bark, next with a hand drill to bore a neat little set of sapsucker holes, last with a plain hammer-driven nail. Nothing. But the month since equinox has not seen rain; the young trees may be conserving their moisture rather than yielding one drop of it to me.

Then serendipity brings another recipe for obtaining a substance very like storax: boil young branches. Before you can say *Liquidambar styraciflua*, a great potful of chopped sweet gum twigs is simmering in water to cover. I may not have storax, but I find the sweet gum's rich and potent fragrance. My decoction tastes woody, like cardboard (or tree), but oh, it smells of cinnamon! No wonder Don Bernal made no mistake.

And as I chop the twigs, I find one with a spontaneous outflow of gum. It does not go into the pot. No special odor rises from the sticky stuff. I decide to test its chewability. It tastes no more delicious than the liquid decoction. The gum stays stuck to my teeth through two brushings.

V

Sweet gum figures twice in Audubon's paintings of birds. The first treats its subjects realistically. It shows Traill's flycatcher, a species that the artist himself discovered in 1822 on the Arkansas River near its confluence with the Mississippi. The little bird perches on the tip of a twig just coming into spring-green leaf. In the customary fashion of sweet-gum twigs, it bears what the botanists call "wings"—several thin but prominent ridges of corky bark along its sides. The second painting is an exercise in artistic license, if not outright fantasy. It depicts birds and sweet gum clearly enough: the tree's mature seedpods dangle brown and prickly beside dark green leaves; the birds in the branches are delineated in all their fine colors. On the page they are quite alive, a boisterous, posturing hodgepodge of crowkind—scrub jay, Steller's jay, Clark's nutcracker, yellow-billed magpie. And there's the trouble: the ranges of these corvids and the sweet gum do not coincide. The sweet gum grows from Connecticut south into Florida, west through Oklahoma to the easternmost part of Texas, and north into southern Illinois. But the birds are, every one of them, species of the far West. And of the four, only the scrub jay—a bird fond of chaparral, scrub oak, and stands of piñon juniper—might ever be found screaming raucously from a sweet gum; an isolated eastern population occurs in the sweet-gum country of central Florida.

But Audubon can't really be faulted. Here, appearances are everything. It's the accuracy of depiction that counts rather than outstanding fidelity to environmental facts. We are meant, anyway, to attend most closely to the birds, not to the vegetation.

Vegetation serves as a backdrop to feathered action in all of Audubon's paintings. So, he often delegated its renderings to various assistants. Sometime in the 1830s, he assigned the task of painting the greenery for crowkind to the skilled hands of Maria Martin. (She later became wife to the Reverend John Bachman of Charleston, South Carolina, for whom Audubon named a sweet-singing sparrow and a now rare warbler. And the connection was even closer, for

Audubon's two sons married Mr. Bachman's daughters.) It was another assistant who, in 1822, painted the perky, new-green leaves and winged twigs for Traill's flycatcher—Joseph Mason, aged fifteen.

VI

This is an opinion held widely throughout the South: the sweet gum is so ordinary that it's downright common. Worse, it's an opportunist, often the first deciduous tree to root itself in brush piles, clearings, or abandoned fields. Turn your back—there's a sweet gum shooting skyward with the brash green insistence of a thistle. It's a trash tree.

Cross the Mason-Dixon line. Head inland or along the coast into southern New England. The farther north, the more reluctant the sweet gum to rise up in any and every available space, and the greater the esteem in which it's held by the human population. Once, when I lived through four months of northern exile in New Jersey, I noticed that sweet gums occupied honored places in the landscaping of many lawns. And on the grounds of the private five-bedroom, five-bath house in which I boarded (low rent in exchange for serving as cook, errand girl, and chauffeur to small, brown, car-crazy dog), two sweet gum trees guarded the outer corners of an ample guest-parking area. A hired lawn-and-tree service vacuumed up untidy seed capsules, sprayed leaves against possible pests, pruned limbs for the sake of shapeliness, and anointed and bound any small wounds (some, I think, imaginary). Pampered trees!

In truth, sweet gum resists most pests and diseases. This tree with the cinnamon fragrance and star-shaped leaves is not only beautiful but also tough.

VII

In April, two weeks after equinox and the sapsucker's vanishment, the sweet gums begin to bloom in our yard and, indeed, all up and down the Carolina coast. New Jersey, land of my exile, won't see such activity for another month.

Blooming: the species is monoecious—"single-housed"—with each tree bearing both male and female blossoms. Each tightly clustered group of male flowers looks something like a miniature Christmas tree—triangular and decorated with tiny, closely set greenish balls. Their female counterparts droop daintily from slender stalks at the triangle's base. And the styles of the female flowers—the style is the part of the pistil between the ovary and the stigma's pollen-receiving platform—are twisted and weirdly curved. As green as a pea at first but much smaller, the seed capsules—John Lawson's "Burs"—won't ripen fully till fall. By then those styles will have turned into interlaced prickles—many two-horned capsules stuck closely together.

They're fearsome to look at but blunt and yielding enough not to injure my exploring fingers. (Bare feet on a summer night are quite another story.) Through winter into the following spring, the pods, big as an English walnut in the shell, dangle from the sweet gum trees like dark brown ornaments.

VIII

The third week in May, a golden bird flies into the green leaves of the sweet gum at river's edge, the very tree that the sapsucker favored. After a minute, the bird flies out.

She soon returns, however, a long, thin ribbon of grass in her bill. She's an orchard oriole obeying spring's recurrent and timeless command: Build your nest. The job of building is hers alone. I know that she's found a mate; she wouldn't be carting in materials otherwise. And the pair is monogamous—at least for this season (their fancies may wander elsewhere next year). Both will defend their chosen territory, but the other tasks of rearing young are divided between them. She will construct the tightly woven basket that will be her nest, and for two weeks, with no time off, she will incubate the eggs, three to seven pale white ovals splotched liberally with purple or brown. He will feed her as she sits, and feed the nestlings, too, until they're strong enough to make it on their own after fledging. And he will stand guard throughout lest mother and birdlings come to harm.

I look for him. He's in the vicinity somewhere, but I do not see him, do not catch one glimpse of his lovely colors—rust-red feathers set off by lustrous black. I hear him, though, as he bugles his brisk, come-and-get-it chow call song and signs off with a hoarse *phew* or a two-syllabled *pt-trrt* that sounds for all the world as if he's pulling the plug on a drain.

Meanwhile, she's busy, carrying in strand after strand of fine grass, wrapping them around twigs to serve as suspension ropes, then weaving a deep, narrow basket and lining it with even finer strands and perhaps a layer of thistledown. She has chosen to build in a dense clump of leaves at the farthest end of the tree's lowest branch, which is still twelve feet above my head. The bottom of the little basket peeps out just below the leaves and rocks like a cradle in the river breeze. At May's end she disappears. Laying has begun.

Now I should be able to spot him bringing in food. I don't. But he finally shows himself several days later as he engages in his other instinct-driven tasks—nest-guarding and territorial defense. How fierce he is! The pair of blue jays that I'd thought were youngsters begging in soft voices turn out to be adults murmuring sweet nothings; they're now going about the business of building their own nest, a rickety-looking affair of sticks placed in a high pine crotch. The male orchard oriole will have none of this. Chattering nonstop, delivering every bit of invective in his rather large repertoire, he flies straight at one of the jays, a David taking on Goliath. It seems that he is kept in the air not by wings but by sheer fury. The next day he repeats these maneuvers—the threats, the feinting, the near strike—at an even larger invader of his chosen space, a common crow. All afternoon he charges and screams. Next day, he puts these defensive actions through a full-scale rerun.

And I know the reason that I failed to see him earlier: I'd looked for the wrong colors. No black and rust-red here; this bird is golden like his mate but with a black bib at chin and throat. He's a first-year male, fledged the summer before; like many other species of oriole, including the flame-bright Baltimore, he will not don his full adult regalia until this summer's end. The strategy may be that mimicking female colors prevents ouster by older, more experienced males and assures his own young chance at reproduction.

At the end of the first week in June, there's no activity at all around the nest. The nest itself still sways within its bower of green leaves, but of the two golden birds I see no sign. Did a night wind off the river rock the fragile basket so hard that the eggs fell out? Did smoke rising from our stone-and-concrete outdoor grill, set twenty feet below the birds' branch (we see that now), choke the pair? Did a hungry snake inch its way to a source of food? Or did the crow and the jays work the egg-eating predations for which they are notorious? We'll never know.

IX

Sweet gum's botanical family is that of the Witch Hazel, the Hamamelidaceae, which means (without any logic that I can discover) "like an apple." Without exception, its members are found in the Americas or Asia. The family's only other North American representative is the eponymous witch hazel, *Hamamelis virginiana,* which has nothing to do with apples and everything to do, at least in popular imagination, with water-divining and working great cures on bruises, sprains, piles, and a host of other pesky afflictions from vertigo to blindness.

Thomas Harriot made the first English mention of witch hazel in his *Briefe and True Report of the New Found Land of Virginia,* published in 1590. ("Virginia" included newly discovered crown lands all the way from Chesapeake Bay down the coast of territory that later became the Carolinas.) Commenting on the land's "naturall inhabitants," he wrote, "They are a people clothed with loose mantles made of Deere skins, & aprons of the same around their middles; all els naked; . . .

having no edge tools or weapons of iron or steel to offend us with all . . . those weapós that they have, are onlie bowes made of Witch hazle, & arrows of reeds. . . ." And with that, the common name for the shrublike American tree was set, not for all time perhaps, but for a good few centuries to come.

Forgivably, Thomas Harriot muddled this tree that has no counterpart in the Old World with two somewhat similar, many-branched English species, a British hazel and the wych elm. And with that, American witch hazel, which grows in the wild only from Maine to Florida and reaches its westernmost limit just beyond the Mississippi, took on the magical powers of its British namesakes. In today's world there are those who would not dream of digging a well without first using a forked rod of witch hazel to locate underground water.

For me the tree's miracles come in other guises: the delicate yellow flowers, with petals like threads, that bloom in October; the seed capsules that hang on the tree for nearly a year—and then spring open with an audible snap, hurling seeds as far as twenty feet.

How on earth is sweet gum, which obeys the seasons and towers like any respectable tree, related to this autumn-blooming, propulsive shrub? By the shape of the seeds, that's how. Both trees bear horned capsules, single for witch hazel, many-clustered for sweet gum.

Sweet gum itself seems to have escaped Thomas Harriot's otherwise observant notice. He did record "Sweete Gummes of divers kinds and other Apothecary drugges," but it was the sticky stuff that caught his attention, not the trees from which it came.

X

August: I visit the sweet gums in which I made woodpecker holes with drill and nail last spring. Storax! It's clear, shiny, slightly tacky to the touch. Gathering a pea-sized ball between finger and thumb, I pop it in my mouth and nibble. The trees have done as they're programmed to do—release a plasma of sap from an open wound. But I should have known better: again, the gummy stuff sticks to my teeth, resisting all brushing for two days.

XI

Throughout her *Tree Book,* first published in 1905, the botanist Julia Rogers told anecdote after anecdote to illustrate the contexts within which people view trees. And here is a story that she brought to the sweet gum (though she used a dialect spelling for the gospel song):

> The sweet gum is probably more closely linked with plantation life in the South than any other tree. It grows in the swamps, and many a slave hugged the slender shaft of a leafy gum tree while he waited all day for the north star to point him the way to freedom. Here the 'possum and the 'coon found similar refuge from hunters and their dogs; and it was a hollow gum tree that old "Nicodemus, the slave" was buried in to be waked in time for the great jubilee! . . . I recall with great vividness an old ex-slave's description and eulogy of the tree, and the song he sang, full of the exaltation his dearly bought freedom always aroused in him—especially the thrilling chorus:
>
> There's a good time coming, 'tis almost here.
> It's been a long time on the way:
> Run and tell Elijah to hurry up, Pomp',
> Meet us at the gum tree down in the swamp;
> Wake Nicodemus today!

Sweet gum has served as a way station, then, on both the body's road to freedom and the soul's path to eternal joy.

XII

In late September, the sweet gum near river's edge gives food and shelter to winged migrants on their way to the tropics. White-eyed vireos flit busily through yellowing leaves; juvenile redstarts fly in and fly out, fanning their tails to show large, semicircular flash spots of gold. Sometimes a black-and-white warbler, looking like a miniature woodpecker, creeps along a limb.

Showing one of many possible permutations, the foliage of this particular tree turns yellow every year. But look at the saplings in the hedgerows, the trees at woods' edge. When autumn comes, their inmost colors are revealed—scarlets, dark blue-reds, and glistening, almost molten purples.

The colors seen in fall have always been present in the leaves, ever since they began to unfold. The yellow pigment, carotene, and the pigment for deep reds and purples, anthocyanin, lie beneath the green chlorophyll. From spring through summer, in the process called photosynthesis, the chlorophyll has used sunlight's energy to convert the carbon dioxide obtained from the air and the mineral-bearing water brought up from the roots into sugar compounds. The green leaves also release water as vapor, a phenomenon known as respiration, which serves as a natural air-conditioning to keep a plant's internal temperatures from rising too high for successful photosynthesis. But with the onset of cooler weather, the tree begins to shut down so that it will not lose moisture during the winter. The connections of leaf stems to tree are sealed off. Deprived of its own supply of water, the chlorophyll disappears. And the underlying colors blaze.

From gold to crimson and near-black, these days the sweet gum shows fall colors in every imaginable hue. (People who live in New Jersey, take note.) The wizards of horticulture have developed many varieties that may be had through garden catalogues: Burgundy, which holds its purplish red leaves into winter; Palo Alto, which bursts into red or bright orange flame. Still another variety was not artificially bred but found in the wilds of North Carolina back in 1930. Called Rotundiloba, it bears leaves that are rounded, not pointed, nor does it set prickly seed capsules but must be propagated through cuttings.

XIII

Sweet gum may be a jaunty, green survivor of an unimaginably ancient forest.

Geologists believe that once upon a time before trees, some 350 to 300 million years ago, the planet's landmass consisted of a single supercontinent, which they have named Pangaea—"All-the-Land." Later, perhaps 200 to 180 million years ago, Pangaea began to break apart, forming two large continents, Laurasia to the north, Gondwanaland to the south. The former eventually gave rise to North America and Eurasia, the latter to South America, Africa, India, and Australia.

The gymnosperms, plants like cycads and conifers that have naked seeds, seem to have developed on both of these continents. But according to the current thinking of many paleobotanists, it was in the tropics of Gondwanaland that the angiosperms, plants with seeds enclosed in ovaries, came into being. From these rose the legions of deciduous trees that would later circle the globe.

In North America's earliest, Laurasian incarnation, however, conifers almost certainly dominated the forests. Nonetheless, by the time another hundred million years had rolled past, and the Cretaceous period with its dinosaurs had come and gone, sweet gum and other broad-leafed trees like magnolia, tulip tree, maple, and oak were occupying the land on equal terms with the conifers. How did they get here from Gondwanaland? They came by the usual seed-dispersal systems of wind and water and by overland migration when tectonic movements joined a piece of one continent to the other. It is thought that the juncture of the future South America with Laurasian North America may have been the means of distribution for species found later in the Americas and Asia but not in Europe.

The paleobotanists further theorize that in fairly recent times, a mere twenty-five million years ago, much of the land in the earth's Northern Hemisphere supported a vast mixed forest, with broad-leafed trees predominant over the cone bearers. Remnants of this great forest survive, enabling researchers to calculate the extent of the original and tally the kinds of trees it supported.

Among these, the conifers are represented by California's redwoods and sequoias; the broadleaves most significantly by magnolias and—yes, of course—sweet gums.

XIV

On a cold, calm, bright blue January morning, my husband fixes a satellite dish, the tiny eighteen-inch kind, to the trunk of the sweet gum by the riverbank—the sapsucker's sweet gum. He strings a cable (to be buried later) between dish and house, then telephones to activate the system. By midafternoon, he's ensconced in his recliner watching old western movies that star Rocky Lane and Lash LaRue.

By midafternoon, the winds at rest, the day still crisp and sunny, I find my entertainment not in anything the dish brings but in the tree itself. In its upper reaches, it accommodates half a dozen highly skilled acrobats. They're upside down, swaying to and fro as they feed on seed capsules. How quick, how agile their twists and spinning turns! How graceful their transfers, one capsule to the next! I'm certain that this circus of yellow-shafted flickers is by far the best show anywhere around.

THE FORTUNES OF SASSAFRAS

■

"We used to gather sassafras roots, my friends and I," says my neighbor Mo. "That was back in the early twenties, when I was a teenager living near Gary, Indiana. We didn't have to dig, though. The soil was sandy, and we just pulled them up. There was a man came around used to buy them. His real work was buying furs, but he took all the roots we had and put cash in our hand."

But, oh, the fortunes of sassafras have fallen on evil times since then. For most of human history, its roots, bark, and leaves have been esteemed for their many benefits. From the earliest days, American Indians used parts of the tree to concoct poultices and other healing remedies. Spanish explorers and British colonists plucked and shipped it back home as the latest thing in a cure for every ailment. Not only physicians and patients but also cooks have found the tree to be a source of quite miraculous ingredient. The leaves have long been dried and crushed to make filé powder, a prime thickener for southern stews and gumbos. For generations, people like Mo and his friends have harvested the roots as a sideline and sold them to commercial buyers. Those roots are a prime source of safrole, the aromatic oil that was used till recently to flavor soft drinks, candy, and medicine. But in 1960, the Food and Drug Administration declared the oil toxic and banned it from interstate commerce. Once respected as a panacea, sassafras is now regarded as a poison. I think it's a bum rap.

Who's right, the FDA or centuries of sassafrassers? Mo, now in his early seventies, wonders about that as he introduces me to the stubborn tree in his back yard at Great Neck Point, on North Carolina's wide and salty river Neuse. Actually, it's not one tree but several surrounding a stump. The stump, a foot high, ten inches in diameter, is weathered silvery grey and riddled with insect holes. The young trees have sprouted from its roots. They're sturdy but slender; only one is too large to be encircled by my finger and thumb. I am reminded of old stories about invincible armies: when one warrior falls, two more rise from his body. In the case of this tree, the effect is even more extraordinary—not just two but an astounding five!

Such, however, is the habit of the tree now formally known as *Sassafras albidum*, or whitish sassafras. It's hard to guess why anyone would call it "whitish"—the bark is reddish brown, the heartwood an orange-brown—but the name may refer to the pale yellow sapwood or the white, mucilaginous pith in its twigs. Some early botanists, noting the tree's great service as an herbal remedy, dubbed it *S. officinale*, "medicinal" sassafras. Others, looking at the shape of things, opted for *S. variifolium*, sassafras "with diverse foliage," because the leaves on any one tree assume three forms—entire, lobed like a mitten with a thumb, and three-lobed with separate compartments for thumb and little finger. By any of these names, the tree is native to North America and widespread east of the Mississippi, especially in the South, where it sneaks across the great river into Texas. And the tree may spread either by its beautiful seeds— berries of shiny midnight blue, each one on its own sunset-red stalk—or by suckers popping up from the roots. Mo's backyard sassafras has taken a lease on longevity through the latter method; its present five trunks would likely give rise to more and become a small thicket if he neglected to mow his yard.

It may be that the parent of the five was male. Sassafras is one of the species that the botanists call dioecious—"two-housed"—with pollen-bearing male flowers on one plant,seed-producing female blossoms on another. But by a marvelous vegetative trick (known to many other kinds of trees), the male sassafras may reproduce without reference to any member of the opposite sex. The tree simply clones itself by sending up suckers and more suckers until it stands amid a small community of genetically identical offspring. In the same way, the female sassafras may generate a grove of daughters precisely like herself. These stands, however, do not occur alone;

S. albidum is sociable, with a special preference for the company of oaks and sweet gums, pawpaws and persimmons.

If the Indians, explorers, and colonists knew of the tree's reproductive habits, they almost certainly paid them no heed, except to be delighted when they came across abundant stands. Sassafras piqued their interest by other kinds of magic. One was medical: throughout its range, from New England to Florida, from the Atlantic seaboard to the Mississippi, sassafras had long been an essential part of the native pharmacopoeia. The tree was unfamiliar, however, to Europeans until the early 1500s. Then the Spaniards, arriving in the land to which Ponce de Léon gave the name La Florida, saw the tree, sniffed its spicy aroma, and learned of its reputation as a cure for every ailment under the sun. Discovering sassafras must have been the next best thing to finding a fountain of youth.

The second magic was economic, and it sprouted from the first like a vigorous sucker: The European explorers found that money grew on trees—at least, on sassafras trees. When the folks back home heard of its curative wonderwork, they not only raised a clamor for the stuff but also opened their purses wide. It was a luxury as precious as gold and jewels, as desirable as cinnamon and silk, and had the added benefit of restoring the user's health. The Spaniards began shipping sassafras products early on, and it was they who christened the tree with the name by which it has been known ever since in the Western world. Whether the Spaniards originated the name is up for dispute. It may be a Spanish transformation—linguistically possible but botanically off the mark—of the Latin *saxifraga,* which means "rock-breaking." Or it may have been an Indian word that the Spaniards adopted, though one of them, a Dr. Monardes writing in 1571, gave the native name as *pauame* and attributed the word *sassafras* to the French. But it's doubtful, I think, that the French had anything to do with the nomenclature. Most likely, one tribe's *pauame* was another's *sassafras.*

But there were fortunes to be made, and the French most assuredly shook the money tree as often as they could. So did the

English, who translated Monardes's work in 1577: ". . . the tree that is brought from Florida, which is called Sassafras." Not much later, its presence on the mid-Atlantic coast was noted by Thomas Harriot, mathematician, astronomer, navigator, and member of the 1585 expedition sponsored by Sir Walter Raleigh that brought the first settlers to Roanoke Island. Including the tree on a list of this newfound land's marketable commodities, Harriot described it this way: "*Sassafras*, called by the inhabitantes *Winauk*, a kind of wood of most pleasant and sweet smell; and of most rare vertues in physick for the cure of many diseases." He also averred that sassafras was far better and of more uses than "the wood which is called Guaiacum." Till sassafras bumped it aside, Guaiacum (also called lignum vitae) was thought to be a surefire cure for syphilis.

Then, in 1602, the explorer Bartholomew Gosnold found sassafras in New England as he sailed his ship, the *Concord*, down the coast from Maine to Narragansett Bay. In the course of his voyage, he not only named such notable geographic features as Cape Cod and Martha's Vineyard but also came across healthy stands of sassafras on an island in a small group, which he called the Elizabeths in honor of the queen, that lie northwest of Martha's Vineyard. (Just which island is a matter of dispute—Naushon, the largest of the group, or Cuttyhunk, where Gosnold built a fort.) The sassafras was harvested, of course, and taken back to England, where its products were sold for a princely three shillings a pound, a sum that amounts to well over a hundred dollars in modern terms. Four years later, commanding the *God Speed*, Gosnold transported some of the first colonists to Virginia. Though his complaints about the site of Jamestown went unheeded and he died there of malaria in 1607, he did not leave the world without observing that Virginia, as well as New England, provided suitable habitat for this tree as good as gold.

Captain John Smith has attested to that fact and to the high value placed on sassafras, which helped in the earliest days to keep the brand-new settlers from starvation. The third book of his *Generall Historie of Virginia, New-England, and the Summer Isles* relates "The

Proceedings and Accidents of the English Colony in Virginia." And accidents certainly happened. When the ships that had landed the colonists set sail for their voyage home and disappeared over the eastern horizon, and the supply ships bringing new provisions had not left port, those lingering on the shores of the New World fell upon hungry times:

> Being thus left to our fortunes, it fortuned that within ten days scarce ten amongst us could either goe, or well stand, such extreme weakness and sickness oppressed us. And thereat none need marvaile, if they consider the cause and reason, which was this; whilst the ships stayed, our allowance was somewhat bettered by a daily portion of Bisket, which the sailors would pilfer to sell, give, or exchange with us for money, Saxafras, furres, or love. But when they departed, there remained neither tavern, beer-house, nor place of relief, but the common Kettell.

And nothing, not sassafras nor love, could then buy food to supplement the dreary contents of that kettle. Each man's daily ration consisted of a pint of wheat, along with barley boiled in water. According to Captain Smith, these victuals, "having fryed some 26. weekes in the ships hold, contained as many wormes as graines; so that we might truely call it rather so much bran than corn, our drink was water, our lodgings Castles in the ayre. . . ."

Meanwhile, back in England, Shakespeare's friend Ben Jonson wrote in his play *Volpone,* first published in 1605, of a quite imaginary medicine, with which the play's protagonist attempts a scam. He says that this nostrum, heretofore a well-kept secret, is so potent that had its existence been earlier revealed, the upstart New World medicines like sassafras wouldn't have had a chance: "No Indian drug had ere been famed, / Tabacco, Sassafras not named. . . ." But from the time that sassafras was first brought to England's attention, it continued to be eagerly sought out and bought as a wonder drug. (Tobacco shared its reputation for working miraculous cures.) Three-quarters of a century after *Volpone,* New World sources for the wonder tree were still being recorded. In 1684, William Penn noted "Sarsafrax" as a species found in Pennsylvania, along with mulberry, ash, and chestnut.

At the beginning of the 1700s, the amazing curative powers of sassafras were still much in demand. Some of the boons of this "Indian drug" were listed with great appreciation by John Lawson. In *A New Voyage to Carolina,* his 1709 account of native inhabitants, geography, natural history, and not-to-be-missed opportunities in colonial real estate, he briefly mentions sassafras's suitability as a material for bowls and houseposts (the heartwood is resistant to decay) and then gives unstinting praise to the medicinal properties found in sundry parts of the plant. Its white flower, eaten in the spring with fresh salad greens, was said to cleanse the blood. Its black, oily mature berry could stimulate the expulsion of intestinal gas and was the prime ingredient of a clyster—an enema—used to relieve colic. The bark of the root was ingested by those who suffered spasmodic intestinal pain. And the root, powdered or made into a lotion, was often applied to reduce a swelling or heal an ulcer.

The book provides Lawson's eyewitness testimony to one procedure using sassafras. As they traveled through South Carolina, one of his party developed a sore knee, a most inconvenient condition for anyone slogging through the wilderness on foot. But as it happened, they met one of the Indians' war captains, "a Man of great Esteem among them," who invited the entire party to pause at his house until the gimpy traveler could again walk without discomfort. So, "entertain'd very courteously" by their host, they settled in for the night. Then, in the morning, the war captain

> desired to see the lame Man's affected Part, to the end he might do something, which (he believ'd) would give him Ease. After he had viewed it accordingly, he pull'd out an Instrument, somewhat like a Comb, which was made of a split Reed, with 15 Teeth of Rattle-Snakes set at much the same distance as in a large Horn-Comb: With these he scratch'd the place where the Lameness chiefly lay, till the Blood came,

> bathing it, both before and after Incision, with warm Water, spurted out of his Mouth. This done, he ran to his Plantation, and got some *Sassafras* Root, (which grows here in great plenty) dry'd it in the Embers, scrap'd off the outward Rind, and having beat it betwixt two Stones, apply'd it to the Part afflicted, binding it up well.

Two days later the knee had healed. John Lawson and party went on their way.

It was just this kind of miracle that had led to Europe's rage for sassafras. And by an odd quirk of fate, the ships bearing spicy cargoes from the New World to the Old were doing far more than transporting a valuable item of commerce to eager consumers. They were bringing *S. albidum* home. The species is abundantly found not only in the fossil records of North America but also in those of Europe and Greenland—records that go back more than a hundred million years into the Cretaceous period. Conifers still ruled among plants then, and dinosaurs walked supreme among the animals. But flowers, mammals, and deciduous trees had begun to appear. On the now tame Carolina coast, where Mo's five little trees are flourishing, sassafras first sprouted recognizably in a world of vegetarian duck-billed dinosaurs and predatory crocodiles fifty feet long. It sprouted and thrived in the company of other newfangled trees with leaves obedient to the seasons—magnolia, sweet gum, tulip tree, oak, maple, poplar, and baldcypress. Since those early days, sassafras's tenure in North America has not been interrupted. The records show that it survived until fairly recent times in Europe and lost its hold there only when the last glacial ice retreated 11,000 years ago. So, it may be that prehistoric Europeans, vanished Neanderthals and ancestral CroMagnons, knew sassafras; that they, too, savored tea made from its steeped roots and used the bark to heal their wounds.

Two other species of sassafras are found in the world, both of them in the Far East. All three belong to the order Laurales and, more particularly, to its most economically important family, the Lauraceae. What joyous, tumbling repetition in these names, Laurales

Lauraceae—Laurels That Really Are Laurels! And the family is characterized by fragrance, spice, flavor, and zest. It holds, as well, the heady scent of victory: the leaves of one of its species, the bay laurel, were used to make crowns for the winners of the Pythian Games, held regularly at Delphi to honor the god Apollo, and second in prestige only to Zeus's great games at Olympia. (The winners of the Olympian contests were crowned with wild olive leaves. Victors in the other important classical games were awarded crowns made of such materials as dry or green parsley.) Though bay leaves are no longer used for such a purpose, their ancient sweetness has not vanished but reappears whenever the title "laureate" is bestowed upon a deserving soul. (Has anyone ever braided the leaves and twigs of sassafras into a crown?)

It's easy to guess that sassafras is related to such richly aromatic members of the Really-Laurel family as the camphor tree of eastern Asia and the cinnamon tree of Sri Lanka and India. Indeed, an American who surveyed one mountainous western portion of the Virginia colony in 1746 wrote in his journal of marking a "Cinamon tree"—most certainly a sassafras—along his route. But the family is huge, comprising some 2,200 species, many of which have no odor at all. One of these is *Persea americana,* native to Central America and better known as the avocado. Fancy that—sassafras kin to the avocado!

But while the latter's reputation for excellence has been sustained, and perhaps plumped up, the former's has not. The fortunes of sassafras declined at a leisurely pace, however, in the first four hundred years after the Spanish explorers found it in Florida. As a money-making proposition, sassafras soon lost its bloom, for it grew in such plenty and was so easily harvested that the price bottomed out. Nonetheless, trade continued brisk because demand for its products stayed high on both sides of the Atlantic. For several centuries, though nobody got rich, sassafras put change in many pockets, including those of Mo and his teenaged friends.

How versatile the sassafras and how generous, a spice for all seasons, a plant long at home in three kingdoms—those of medicine, commerce, and domestic life. Indians chewed the roots as people nowadays chew gum. Cooks have long relied on the crushed leaves to thicken soups and stews. And thirsty souls have ever enjoyed the tea for its own sake. During the Civil War, it provided Americans of both Northern and Southern persuasions with a fine substitute for the unavailable Oriental product. In the gold-rush encampments of the mid- to late 1800s, "cinnamon root" was used, along with other cover-up coloring and flavoring ingredients like plug tobacco and burnt sugar, to disguise the taste of raw alcohol, which saloonkeepers could then sell as whisky or even as cognac to miners with a thirst. Infusions of sassafras have also been sipped for more benign purposes: as a spring tonic to purify the blood or as a stimulant for the heart. Recipes for sassafras-based medicinal cordials may be found in eighteenth- and nineteenth-century cookbooks. One such receipt (as it would have been called in those days) instructs that the following hodgepodge be added to the basic tea: candied citron and lemon, lemon syrup, sugar candy, muscatel raisins, pistachio nuts, juniper berries, rosemary and marjoram, gum arabic, sarsaparilla, distilled grape spirits, white wine, and sherry. (Imbibing such a cordial would likely have stimulated giggles and staggers as well as the heart.) Sassafras oil has given its flavor to candy and medicine, its perfume to soap. And it has served as the prime ingredient, sine qua non, for root beer.

My own discovery of sassafras—the enticing scent and the drink-me flavor—occurred in the mid-1960s, not long after the fortunes of the tree had fortuned its condemnation as worse than unwholesome and its placement, along with other eat-me-nots like saccharin, on a list of forbidden products. But I knew nothing of such doings, not at that time nor for decades after. My children, then in the elementary grades or junior high school, found young sassafras shoots growing here, there, everywhere in the dappled shade of our two-acre Connecticut woods. And my third child, the daughter who could cast seeds on a stone and without fail watch them grow, pulled up some shoots, caught a whiff of their spicy scent, and brought them in. After we'd washed clinging soil off the roots, we steeped them in hot water. As the water cooled, it began to assume the glowing red-amber richness of fine pekoe, and it smelled like a host of spices—cinnamon and cloves, nutmeg and allspice—had married and merged to become a single, irresistible enchantment. We drank it with sugar and without. We kept it on hand in a refrigerator jug. Year on year, I made sassafras jelly, which we gave away at Christmas or kept for ourselves to slather on such things as English muffins and roast pork.

Enter the well-meaning people who would save us from ourselves. (Surely it would be easier to stay the tides.) Sometime toward the end of the 1950s, someone decided to test sassafras and, more particularly, safrole—the tree's aromatic and highly volatile oil—for carcinogenicity. In a handful of tests, safrole was administered to laboratory rats. The doses were massive, as much as .52 percent of the animals' daily diet. The results of these tests were not surprising: on a sub-acute level, safrole is toxic to rats and may cause changes in the liver.

Why test—why conceive of testing—in the first place? I've asked those questions outright. The FDA has responded with information about the "experiments," as it terms the tests. It has provided material on the chemical makeup of safrole and the analytical method used for detecting this compound. It has furnished data on the rats: the groups used for the experiments were small, from eleven to sixteen animals, about evenly divided into males and females. But it has said nothing—not one single word—on the whys of the matter. It may be (I somehow doubt it) that workers who handled sassafras for commercial purposes showed an unusually high incidence of tumors. Or (more likely) testing was prodded by an excess of zeal, the reasons for which no one recalls—or wants to look up—more than three decades later.

Whatever the rationale for the tests, the FDA behaved predictably when it received the findings. In an order published on December 3, 1960, it announced that safrole is a "weak hepatic carcinogen," and it promptly banned all food containing sassafras extracts or bark from interstate commerce—unless, like decaffeinated coffee, the tea has been rid of its safrole. But even after such purification, the desafrolated extract continues to be governed by federal regulations, which list the conditions under which it may be sold and used.

Well, sassafras! (Makes a pretty fair cuss word, doesn't it?) Such protective interference strikes me as unfortunate on several counts. Sassafras was never, to be sure, the hoped-for panacea. But its current status as a poison seems open to question. Not least of the reasons for disputing the test results is that the methods used to attain them are suspect. Any creature—rodent or piscine, avian or human—will suffer harm from huge overdoses of anything, including substances like vitamins that are beneficial in small amounts.

Then, of Ben Jonson's two "Indian drugs," it's likely that proportionately more users—many more—have survived exposure to sassafras than to tobacco. Nor does the use of sassafras in any form lead to a can't-quit addiction. (Having already confessed to indulgence in sassafras, I confess here to forty-three years with tobacco, a fondly remembered friendship that ended just six years ago.) Yet, tobacco is legal while sassafras is not. But no need to dwell on that irony.

Could be, though, that I've done sassafras a great wrong by mentioning an official ban that seems to be a well-kept secret. In my eagerness to know a spicy old acquaintance even better, I've rummaged through many volumes, from botanical texts and encyclopedias to field guides and cookbooks, all of them easy to come by and a good half of them published after the FDA had issued its warnings and regulations well over thirty years ago. Even so, I might have missed the news that sassafras tea had been tried, found guilty, and labeled with a skull and crossbones: only two of the books I looked at, a pitiful two out of dozens, bother to mention the FDA's findings. Why the large silence? It may be that the FDA has not sufficiently publicized the dangers imputed to the plant; it may be that the dangers have been intentionally discounted by people possessed of good sense. But most likely, it's simply that no one pays much attention to sassafras these days. The tree retains commercial importance mainly as the source of filé powder, which contains no safrole and is exempt, therefore, from the FDA's zeal. But how many people use filé?

How many people remember sassafras? Nowadays, a bushel of the stuff couldn't buy a thing, not even a ship's Bisket. The tree with the tingling aroma and magically spicy taste has been so stripped of all value and enchantment that it's not a promising candidate for rehabilitation. Saccharin, similarly abused by ardent researchers and suffered by overdosed rats, was first banned but later reevaluated and returned to supermarket shelves with a cautionary label declaring in capital letters (albeit very small ones) that the substance "HAS BEEN DETERMINED TO CAUSE CANCER IN LABORATORY ANIMALS." Buyer beware, but the choice is the buyer's, and demand for saccharin runs strong. No market forces, however, support the cause of sassafras; these days, hardly anyone craves the stuff. And it's not only out of fashion but also relegated to a corner so far removed from public notice and desire that lobbying the FDA on its behalf would not be worth an ounce of effort.

But that's all right. Unlike saccharin, sassafras has a life of its own. It grows in the wild, not in some factory retort. No rules of human devising can possibly regulate its natural abundance or prohibit access to its products. Money can't buy it, nor love, but so what? Sassafras is here for the taking, free to anyone who wants to gather up the roots and leaves.

In our time, however, it fortunes that the brewing and savoring of sassafras has become an exclusive private preserve, restricted to a devoted few. The club is not closed, though, and would certainly welcome newcomers who'd like to try spicing their days with a dash of gustatory adventure. And let me say here that in advocating the cause of a tree, I am far from advocating the assumption of a fatal risk. I speak instead for common sense. That good and honorable faculty, along with the official turnaround on saccharin, supposes most reasonably that there's only one way a human being might trigger safrole's effects (if any) on the liver, and that's to drink gallon on daily gallon of sassafras tea. The greater dangers are bursting or drowning.

For love of sassafras, I'll gather roots again, gently wash away the clinging earth, and brew strong red-gold tea. To that, I'll add sugar, pectin, and fresh lemon juice. Mo is on this year's Christmas list for sassafras jelly.

THE FLAVORS OF SASSAFRAS

The cautious, who prefer to abide by official pronouncements, might like to make safrole-free filé powder. Others, who find caution unnecessary, can take a fling at cooking up some full-strength sassafras jelly.

The first requirement for people of either inclination is, of course, to locate a sassafras patch. After that, the cautious have it easy, for leaves are the source of filé, and only two pieces of equipment are needed on the spot: pruning shears and a paper bag to stuff with clipped-off leafy branches and twigs. The bold are in for a bit more work. It's the roots that we need—roots not always amenable to being grubbed up. If the soil is loose and moist, fine; just grab a sapling and pull. If not, be prepared. Along with the mandatory paper bag to hold the take, gathering calls for a four-pronged potato rake for scratching around, a spade for serious digging, and hawksbill clippers or a hatchet for cutting through the thicker roots.

Filé Powder

Filé, an excellent thickener and source of a subtly spicy flavor, is perhaps best known as a characteristic ingredient of Southern, and particularly Cajun, cooking. (One Cajun cook, however, has recently poohpoohed it, saying that it does thicken the broth but adds nothing whatsoever to its taste. I suspect that his tongue has suffered a burnout induced by too many hot peppers.) Nonetheless, filé powder is not easy to come by. In most parts of the U.S. its habitat is the specialty shop, not the supermarket shelf. But those of us who want it may go afield, gather sassafras leaves, and make it ourselves.

When should the leaves be picked? When they're young, say some filé makers, while others advocate waiting till summer's end. I do my own gathering in May. But there's no firm rule. The season for picking is rightly determined by access to an abundance of fully formed leaves. And when the branches have been clipped and brought home in a bulging paper bag or two, what then?

- The leaves, stripped from branches and twigs, may be dried in a dehydrator or a microwave oven. Fear not, however, if neither of these appliances is available. My husband, the Chief, offers another fine method: in warm weather, place the leaves in cardboard boxes; then keep the boxes on the seat of a closed car. They'll dry in a week or so. When the leaves are ready, they will crumble easily when rubbed between the fingers.
- Crush the dried leaves by hand, picking out stems and the larger veins as you go, or put them in a bag and pound them with a mallet. Rub these coarsely crushed leaves through a strainer to create finer leaf flakes and to remove any remaining stems and veins. The fine flakes may be stored as is and used instead of powder.
- To reduce these flakes to a powder, rub them through a tea strainer. (This maneuver takes much elbow grease.)
- Store the powder in a jar with a tight-fitting lid. I use clean spice jars, especially those with shaker tops beneath the lid.

Add filé powder (or flakes) to soups and stews immediately after the dish has been cooked and the heat turned off. Beware: if the heat is still on when the powder is stirred in, it will quickly illustrate its name—which harks back to the Latin *filum,* meaning "thread" or "string"—by forming thready clumps and stringy strands.

Use filé in sprinkles or dollops according to the amount of thickening and the taste desired. It's as good with canned soups—chicken, vegetable, beef—as with the homemade kind.

Sassafras Tea

As A precedes B, tea comes before jelly.

1. Wash the dirt off the roots with cold water. Scrub them gently with a vegetable brush. Use them fresh or dry. Store the excess, when dry, in a canister or glass jar.
2. To make tea, take enough roots to carpet the bottom of a 5-quart pot. A hearty handful should suffice. Snip the thin roots into 1/4-inch pieces. Whittle the thicker roots into shavings.
3. Cover the roots with 3–4 quarts water. Bring to a boil and immediately reduce the heat. Simmer for 15–20 minutes. Remove from heat and let cool.
4. Strain the tea through cheesecloth into a container. Then refrigerate—but not before indulging in a drink, warmed or over ice, with sugar and lemon juice to taste.

Sassafras Jelly

Not a drop for making jelly may be left. But if there is, the transformation of liquid into semisolid is quickly accomplished.

Ingredients

3 cups strong sassafras tea
1 package (1 3/4 ounces) powdered pectin
1 tablespoon lemon juice, fresh or bottled
4 cups sugar

- Follow instructions that come with the pectin for timing and adding sugar.

  Fills 5 half-pint jars. Put any leftover liquid in a refrigerator jar for use as soon as it sets.

HOGOO

An Interlude in the Back Field

■

"No! Rabbit tobacco? You don't!" Juanita exclaims. She shakes her head vigorously, and her brown face assumes a wide-eyed look of disbelief.

We say that, oh yes, we do have some, and right here, too. The back field is full of it.

"No!" she repeats and bats her eyes at my husband, the Chief. He grins and invites her to see for herself.

Juanita, in her energetic and flirtatious seventies, is a practical nurse by trade and a keeper of traditions by birth and inclination. It was eight years ago that she first visited Great Neck Point on the east bank of the river Neuse. She was caring then for the elderly mother of the neighbor who owns the weekend cottage immediately downstream from our place. When the mother was brought out from town for a day of sunshine and fresh air and eating fish caught that very morning, Juanita came along. And after the first visit, she never arrived without her fishing pole, the old-fashioned kind made of cane that country women have always used. To this day, such ladies may be seen sitting or standing on ditch banks or by the sides of ponds. They often look like ladies, too, wearing calf-length dresses and wide-brimmed hats to keep off the sun. But when it comes to fishing, they're all business, wielding those limber poles with the utmost efficiency and hauling in buckets of catfish, perch, bream, and sunnies. Here on the river, Juanita usually fills her bucket—a capacious five-gallon pail that used to hold joint compound—with the plumpest pinfish I've ever seen. When our neighbor's mother died late in her ninth decade, Juanita's caregiving job came to an end, but not her visits to the river.

But Juanita's not here today for the fishing. Early April, the water still holds a wintry chill, and a wind out of the north is kicking up whitecaps—hardly the right conditions for catching pinfish.

Juanita's been propelled all the way from town this time by an urge to show off her new Easter bonnet. With a stiff, upturned brim, it's as jaunty as a sailor's hat, but larger and made of a boldly checked red-and-white fabric. She puts it atop her grey curls at a saucy angle and grins. We all duly admire not only the hat but her high style. It's easy to understand her allure for what she calls her "boyfriends," one of whom, a courtly and handsome preacher, accompanied her recently to the river.

"Nobody at church gone have a hat like this Easter Sunday. It is one of a *kind,*" she says proudly and glides without pause into the next subject on her mind. "That sassafras you want—bring it to you real soon, hear? Sassafras tea, now *that* will clean the blood."

She drinks it for that purpose, as she's done every spring since she was a small girl and her mother simmered up gallons of the purifying tea. When Juanita recently acquired a large sassafras tree, cut down in the woods, she had our neighbor put it through his wood chipper. I want some of the fine-chopped bark and roots, all right, but not for brewing spring tonic. The tea is the base for a delicious jelly, and I tell her so.

"Jelly! Never did hear of that," she says. "But let me tell you something else my mother cooked up out of things just grow in the wild."

And that's how we arrive at the subject of rabbit tobacco. Her mother would boil its leaves, strain off the liquid, and use it to bring down a fever. "My brother and me, we called rabbit tobacco 'life everlasting' because it was good for so many things. A lot of medicine grow out there in the wild. People don't know it these days. They think rabbit tobacco just something for little boys."

"Little girls, too," the Chief says and asks if she ever smoked it when she was a child.

"Hunh!" she snorts, tossing her Easter-bonneted head, and proceeds to regale us with accounts of other herbal medicines used to miraculous effect in her growing-up days.

Later, the Chief tells me that if he'd known way back when that rabbit tobacco was a cure for anything, he might not have smoked it.

Then he backs off. "No, that's wrong. My friends and I, we would have smoked the stuff no matter what. Smoking was the adult thing to do, but we couldn't afford buying cigarettes. First time I tried it, I was in second grade."

It would have been the spring of the year, probably April. It's then that rabbit tobacco often grows thick in the fallow fields of eastern North Carolina. It pops up a foot or so high at the end of colder weather and seizes its chance at light before the more rank and rapacious weeds—docks, smartweeds, sow thistles—crowd it out. Its formal name speaks of its appearance—*Gnaphalium obtusifolium,* cottonweed with blunt leaves. Fine, cottony white hairs frost the stems and lend a shimmer of silver to the undersurfaces of leaves, tiny stemless leaves not much bigger than my little fingernail. The plant's common names are legion. Not only is it called life everlasting for its supposed curative powers but also catfoot for the touch-inviting softness of the hairs. It's known as poverty weed for its preferred habitat in old, dry fields, and as sweet balsam for its fragrance. The dried leaves do smell as sweet as newly mown grass on a summer evening. Still another common name is cudweed for—I don't know what until Floyd tells me, but more about that later.

"It's the leaves that you smoke. You wait till the whole plant gets old and turns white. The leaves are dry then," says the Chief. And he tells me that they were pulled off the plant, rolled in newspaper, and put to the flame. As a wrapper for rabbit tobacco, the front page of the Whiteville, North Carolina, *News-Reporter* was most desirable because it had a lot of uninked space around the masthead, but sometimes toilet paper was used as a substitute. Forget the Sears Roebuck catalogue; its colored inks tasted horrible.

The next thing I know, I'm making an informal survey of the community here at Great Neck Point. The question posed to friends and neighbors is, of course, What did you smoke when you were a kid?

The question assumes that just about everyone has experimented with smoking. The assumption is not unwarranted in this part of the world, where people—old and young—who smoke and chew tobacco, who dip snuff, may still outnumber those who don't. Bright leaf tobacco has long been a prime cash crop here, and square tobacco barns stand clustered at the edges of the fields. The barns, no longer used in curing, have become dilapidated, but time was that their rafters and floors were covered with crisp little curls of golden leaf. Youngsters rolled those scraps in newspaper wrappers or stuffed them into corncob pipes (and got a paddle or a razor strap whacked across their tender bottoms when they were caught). A lifelong fondness for the leaf often comes early to those brought up in tobacco country. Nor does it matter a whit that in the olden days—the teens, twenties, and thirties, that is, decades before a surgeon general saw fit to issue warnings—people knew full well that cigarettes were coffin nails and that the weed in any form was filthy.

Indeed, back in Sir Walter Raleigh's day, when tobacco was introduced throughout Europe, the drawbacks of smoking were pointed out by quite a few courageous souls who were willing to buck popular opinion. In those years, tobacco was widely held to be the wonder drug of the century, if not the millennium. The 1597 edition of *The Herbal, or General History of Plants,* by the Elizabethan botanist John Gerard, lists tobacco's "Vertues" from A to Z and on through a second round of A and B; the leaves—dried, boiled into tea, mixed with turpentine and wax, mashed into a poultice—were variously said to cure "Migraime" and agues, to stave off hunger pangs and drive worms from the belly, to ameliorate the itching of piles, reduce flatulence, and heal both simple cuts and bone-deep wounds. *The Herbal* also reports, under Vertue Q, that "the priests and Inchanters of the hot countries do take the fume thereof untill they be drunke, that after they have lien for dead three or foure houres, they may tell the people what wonders . . . they have seen, and so give them a propheticall direction. . . ." Miracles! But among those vigorously resisting such beliefs, among those who railed and preached against the weed was James I, perhaps most notable nowadays for the translation of the Bible that he authorized. In 1604, the good king took up his pen to write a polemic. He wrote it in Latin, but here is a contemporary translation of the regal words:

> A custome lothsome to the eye, hatefull to the Nose, harmefull to the brain, dangerous to the lungs, and in the black stinking fume thereof, nearest resembling the horrible Stigian smoke of the pit that is bottomless.

Smoking—what a peculiar act! How did anyone ever think it up?

As I make the rounds of the neighborhood, putting my question to all and sundry, I speculate about the genesis of smoking—of igniting vegetable matter, intentionally drawing the resultant products of combustion into the mouth and lungs, and blowing them out again in a cloud, a stream, or a series of rings. This much is probable: the initial inhalation took place in the New World. Wild tobacco is native to the Americas, and so are the people, the Indians, who discovered its smokability. The plant is a member of the Solanaceae, the Nightshade family, many of which—belladonna, jimsonweed, and datura, for three—contain alkaloids with narcotic properties. Tobacco's notorious alkaloid is nicotine, a substance also found in well-respected Solanaceae like tomatoes and eggplant. (Obviously, to smoke nicotine is one thing; to fry, broil, or stew nicotine and gobble it up, quite another.) At least one species of the wild tobacco smoked in the Americas was potent enough to cause unconsciousness—those "priests and Inchanters" who partook and afterward lay as if dead. The earliest use of tobacco seems not to have been recreational but rather reserved for ceremonial occasions, like peyote among western tribes (who not only ate but also smoked these cactus buttons).

But that first snootful of smoke—how did it happen? Why was the act repeated? In the absence of hard facts, imagination kicks in with its own scenario. This much is possible—that naturally dried tobacco leaves somehow caught fire; that passersby breathed in the strong fumes and experienced something that modern law-enforcement authorities might describe as impairment—a giddy, fall-down, perhaps hallucinatory inebriation. That stuff could knock your moccasins off. Power of this sort came only from the gods and might be invoked only in their service. Tobacco itself was holy. But I'll bet that its connection to the sacred had become attenuated and that use of the leaf had spread from strictly sacramental occasions into daily life well before the early 1500s, when the Spanish explorers came across the tobacco-smoking Indians of Mexico, Hispaniola, and La Florida, and certainly before 1584, when Sir Walter Raleigh's expeditionary adventurers found the Indians of the Carolina coast "drinking smoke." And Native Americans long ago created blends that reduced tobacco's knockout punch. Among the many botanicals they used as additives were the seeds of water hemlock, the dried leaves of cranberry or sumac, and the inner bark of dogwood; the last three of these often figured as ingredients in the tobacco mixtures that Algonquian-speaking peoples called kinnikinnick. (Sixteenth-century diarists also observed that Indians smoked many vegetable products plain, without tobacco, and that they could become as drunk on sumac—and rabbit tobacco?—as on the potent weed.) From personal acquaintance with nicotine's delights, I'm sure that Native Americans puffed away whenever the craving took hold. Nor did they stop at puffing but also chewed wadded leaves or ground them up for snuff.

John Lawson, gentleman surveyor of North Carolina, took note of the craving in his 1709 description of Indian lives and cultures and gave a useful, though sadly forgotten, word for the element in tobacco that lures the smoker on. Of the Indians, he says:

> Their Teeth are yellow with Smoaking Tobacco, which both Men and Women are much addicted to. They tell us that they had Tobacco amongst them before the *Europeans* made any Discovery of the Continent. It differs in the Leaf from the sweet-scented and Oroonoko, which are the Plants we raise and cultivate in *America*. Theirs differs likewise much in the Smell, when green, from our Tobacco, before cured. They do not use the same way to cure it as

> we do; and therefore, the Difference must be very considerable in Taste; for all Men (that know Tobacco) must allow, that it is the Ordering thereof which gives a Hogoo to that Weed, rather than any Natural Relish it possesses, when green.

Hogoo: the word, a mid-seventeenth-century Englishing of the French term *haut goût,* referred first of all to a high or piquant flavor, a relish to titillate the nose and please the palate. The word had a dark side, too, and was sometimes used of foul smells and tastes—meat gone "high," overripe armpits, smoldering sulphur. But before another hundred years had passed, it slipped from notice. Pity. It deserves far better.

Hogoo: the word encapsulates not only flavor, zest, and yellow teeth but also hoodoo and profanity. Hogoo means pigginess and stashing tobacco around the house, just as an alcoholic stashes bottles, so that there's not the slightest risk of running short. It conjures up the knock-your-mocs-off rush just after the morning's first puff. It summons Humphrey Bogart, who, even as a ghost, has a cigarette firmly attached to his lower lip. And Hogoo must have been the powerful magic that inspired British colonists to devote so much time and land to growing tobacco that they had nothing, not energy nor acreage, left for raising their own food. With an almost audible dismay, Lawson wrote that the settlers had to depend on supply ships to bring in their daily sustenance.

While tobacco may not have been the first plant put in a pipe and smoked, it was certainly the most prominent. And it's tobacco that's now grown and used around the world. This state of affairs can be attributed directly to the European explorers, who picked up the habit (without its sacred connections) from Native Americans in the sixteenth and seventeenth centuries, brought leaves and seeds home, and spread the practice of smoking wherever they went. The smoking of vegetable matter other than tobacco—hashish, opium, and the like—wasn't even thought of until the Europeans came along, puffing on their pipes and savoring the Hogoo in brown-gold leaves.

But tobacco is another story, and one not infrequently told. The matter at hand is smokables that lack Hogoo—rabbit tobacco and other such kid stuff. My survey of friends and neighbors proceeds apace: What is the first thing you ever smoked?

I catch my seventy-two-year-old neighbor, Mo, when I go to visit the fruit-laden pawpaw trees in his back yard. He's outside gathering ripe prune-plums. Ripe and unripe, what glorious colors they show—rose, dark red, midnight purple! Considering my question, Mo ceases picking and draws meditatively on his pipe, a store-bought corncob filled with fragrant Carter Hall tobacco. Then he clamps his teeth around its stem and says, "Cigarettes."

No kid stuff?

"No, I didn't get started until I was twenty—six years old and in the service."

But Floyd, who lives in the small grey-green cottage catty-cornered from Mo's place, ambles over to see if he can cadge some plums. Now a year or two shy of collecting Social Security, he's spent his life on this Carolina coast. And when he hears the question, his response is immediate. "Rabbit tobacco—that's what me and my friends used. Chewed it, too."

Aha! Floyd has just explained the name "cudweed." I can see him and his friends with cheeks stuffed full, like hamsters or baseball players. But when I ask if they enjoyed this cud, he says in a don't-be-silly tone of voice, "Hell, no."

I put the question to Al, recently retired sheet metal mechanic and current captain of the volunteer fire department's local substation. He's at work in his garden two lots downriver from our place, but he leaves off mulching the tomatoes (nightshades, full of nicotine) and stubs out his cigarette before he replies. "We pretty well smoked the stuff we had—corn silk, clover, alfalfa. Always some alfalfa in the barn," he says.

I should have expected that. Al grew up on a farm in Indiana.

He continues. "Made our own corncob pipes, too. Drilled a little hole in the side, stuck an elderberry branch in there for a stem.

Branch would be about as big around as a pencil. We'd hollow it out with a wire."

Another neighbor also admits to corn silk. "Yeah, we did that when we could get it. Trouble was, we didn't get it very much," Kenneth says in a Third Avenue accent. New York born and bred, he spent his working days as an engineer driving commuter trains for the Long Island Railroad. At eighty-three, he's trim and white haired, with eyes of purest cornflower blue. For the last seven years, he and his wife, Anne, have abandoned their winter home in Florida to summer at Great Neck Point. Both are what might be called recovering smokers. But though he gave up regular use several years ago, Kenneth is given to occasional bumming.

"But you don't carry them anymore," he says to me, shaking his head. Then he's off on a round of reminiscence. "I sometimes smoked cigarettes when I was a boy—*real* cigarettes. I'd pick up butts and get what was left of the tobacco out, roll it in paper—"

"The paper, newspaper, magazines—my friends and I smoked the paper," Anne says. Her eyes are as blue as his. "No tobacco, just the paper rolled up real tight."

"I tried that, too. Tasted terrible," he says. "Let me tell you how we made our own pipes. Not corncobs. Acorns. Cut off the tops, dug out the meat, and stuck a straw—a drinking straw—into a hole we drilled in the side. Damn acorn would flop around and dump everything out."

And he looks at me wistfully. "If you had a cigarette on you . . ."

"Now, Kenneth," says his wife.

"Acorn pipe, oh yes," says another recovering smoker, our seafood-dealing friend Cap'n Harry, who lives around the corner on the downriver side of Great Neck Point. "Acorn pipe with a reed for the stem, but I never put tobacco in it."

He put something else in it, though, and the something was dried and crumbled oak leaves. But he preferred smoking oak root. Oak root? Yes, and more precisely, the pithy cores of the roots of live oak trees and water oaks. Cap'n Harry grew up on one of North

Carolina's barrier islands before a bridge connected it to the mainland. Out there in sandy, salt-encrusted isolation, he and his cronies would break off pieces of the roots that sprawled along the surface of the ground and had been cured to perfect dryness by the wind and sun. They popped those gnarly lengths into their mouths like great cigars. Cap'n Harry stopped smoking oak root when he was about ten. "My mama caught me," he says, "and let me tell you, she whupped my bottom good."

"Oak root? Well, I never. It was rabbit tobacco we mainly used," says our next-door neighbor Tom, whose mother-in-law was cared for by Juanita of the cane pole, Easter bonnet, and saucy tongue. Tom has the fisherman's equivalent of the gardener's green thumb; flounder, Spanish mackerel, and bluefish fairly leap into his nets when nobody else, not even Juanita, is catching so much as a pinfish. Though never a serious smoker in all his sixty-odd years, Tom nonetheless conducted the predictable experiments as he headed toward adolescence. "But when we could get it—mind you, those were Depression days—we went for coffee, ground coffee."

I canvass a member of a later generation. "Cigarettes," says slim, blonde Linda, lighting a cigarette. She and Tom's red-bearded, swashbuckling, piratical son have just built their house in the field behind Tom's cottage. Then Linda murmurs, "Um," as if she has something more to say but is not quite sure that she ought to say it.

I won't let her get away with that. " 'Um'—would you care to expand on that statement?"

"Well, um . . ." The truth takes a minute or two to emerge. "Fungus."

And not just any old fungus, but the shelf fungi, the polypores and pleurotus mushrooms, that grow on tree trunks. At the end of the 1960s, when she was fourteen, she and her friends plucked them. The thicker end was fashioned into a pipe, with a bowl carved out in the top, and a smaller, connecting hole drilled in from the side. They'd fill the bowl with crumbled fungus. No stem was used. They'd touch a match to the dry bits and simply inhale through the hole.

How on earth did they think of smoking mushrooms? That was the heyday of the flower children. Were Linda's fungi part and parcel of that era's emphasis on psychedelic experiences?

"No, we were good kids. It was a back-to-nature thing." She looks at me quizzically. "And what did you smoke?"

"Monkey cigars, I bet it was monkey cigars," says the Chief, who's been listening in.

I know what he means by monkey cigars: catalpa beans, the cylindrical dark-brown seedpods of the catalpa tree. But I tell him sternly that I wouldn't know how to smoke such a thing. If there's a cigar monkey present, it can be no other than the Chief himself. Turns out I'm right.

"Take the pod. Break off both ends," he says. "Light the end you don't put in your mouth."

"C'mon," Linda coaxes. "You must have smoked something."

Well, yes. Hollyhock stalks.

I explain that the common garden hollyhock's dried stalks, as firm and round as stogies, contain a pith that can be made to smolder by the diligent application of many matches and enough puffing to make the smoker as dizzy as an Inchanter (though I certainly didn't know about Inchanters then). Come late summer, the stalks were available everywhere, not just from garden plants but from those escaped to roadsides and fields. Which member of my klatsch of preadolescent females thought this one up, I do not know. What I do know is that I've never heard of any others who've indulged in the smoking of dried hollyhock stalks.

There the survey ends, though rumors of still other fume-producing substances keep drifting my way: cattails—"punks"—put in a pipe, grapevines smoked like cigars, dried mullein or grape leaves crumbled and rolled in newspaper. A scrap of memory tells me that tobacco companies have also manufactured cigarettes from leafy substances other than tobacco—lettuce, for one—but when I try to confirm this vague but persistent recollection, not one of the four tobacco companies I contact bothers to reply. Perhaps the companies are simply lying low, avoiding any suggestion that leaves other than tobacco are worth smoking. I think it more likely, given the current climate of opprobrium for cigarettes and those who smoke them, that they're running scared. To reply to a stranger's inquiry might be foolish. After all, admitting to the one-time existence of such sideline products could be tantamount to admitting that tobacco is not such great stuff as its proponents, from John Gerard to R. J. Reynolds, would have us believe. And—a corollary—it might be assumed that the companies thought anything, even lettuce, would do in a pinch to keep a health-conscious user satisfied and smoking. But market forces long ago did in the lettuce gambit.

I'm satisfied by my survey's results. But just like a cigarette tapped from a pack, another question pops up. From rabbit tobacco and oak root to tight-rolled paper, coffee grounds, and cattails, what were we after?

Hogoo, that's what I think. This higgledy-piggledy assortment of stand-ins for tobacco has Hogoo, after all. John Lawson used the word as a term for something quantifiable, for a flavor—a "Natural Relish"—that could be achieved only when the tobacco leaf was cured in a certain British way. Just as surely, though, for all that Lawson scorned the taste of their tobacco, the Indians would have claimed Hogoo (had they known the word) to be inherent in their own leaf, both green and cured.

But natural relish didn't figure in our game. All of our kid stuff, every last leaf and stalk and twist of it, tasted terrible; it was worse than bad meat, armpits, and sulphur all rolled into one. No matter what botanical we chose to ignite and suck on, the flavors were uniformly strong and harsh; the reek, the fume nothing other than stygian. We coughed and we retched. No nicotine high knocked our socks off. Yet, till somebody caught us, we kept on packing pipes, rolling weeds in various papers, and puffing away on stalks and roots.

Of course we were after Hogoo! That certitude fits into my mind the way the warm, smooth bowl of a pipe fits into a hand. Our Hogoo was simply a youthful version of the Lawsonian ideal—an ideal defined not by reason but by an incontrovertible logic of the senses. Hogoo in any form is hardly an intellectual construct but rather a sensual phenomenon, perceived by nose, tongue, and muscles; by emotions; and probably by instinct. Hogoo is taste, even one so vile and loathsome that it proves every other flavor ambrosial. It is the sulphurous smell of a match, the searing heat of indrawn smoke, the freeway smog of exhalation. It's a dizziness, too, from the rapid puff-puff-puffing that keeps the pipe or root alive with glowing fire. It's monkey-see, monkey-do, an aping of the elders, of the movements and attitudes that seem to confirm adulthood—that desirable but infinitely distant state. This imitation of adults is intimately conjoined, of course, with the thrilling dreadfulness of disobeying them.

Most of us did go on to smoke tobacco; fooling around with kid stuff is an almost infallible sign that there'll be an association later on with pipes and cigarettes. And it was then that we found another order of Hogoo. Along with a nicotine-induced narcosis, we granted ourselves permission to flirt with death in the form of those foul-smelling, evil-tasting coffin nails we'd always known about. Yet at the same time that we actively, pleasurably courted our own demise, we also found a special comfort—the instinctive solace of a thumb in the mouth.

What did Juanita smoke when she was coming up? (And why am I willing to bet that she put something in her mouth, lit it, and took a big drag? Simply that hers, like mine, was a generation wreathed in smoke.) She dodged the question at Easter time, implying that only little boys indulged themselves in rabbit tobacco and other unseemly combustibles. In hot July, my chance to ask her again arrives with the promised sassafras. She's brought two 33-gallon trash bags that bulge at odd and grotesque angles with roots, twigs, sawed-off branches, and tree trunks split into quarters. It's more sassafras than I can brew in a lifetime. When we've unloaded the bags, spread the damp chunks and logs on the deck to dry, and seated ourselves in the shade, I try to find out just what kind of Hogoo she got into when she was a girl.

She doesn't speak. Instead, she transforms herself. Her nose begins to twitch, her lips to move as if she were nibbling at leaves, small silvery leaves on a slender stalk. If she were to rise and hop away, I'd glimpse a tail as white and fluffy as a powder puff.

"What rabbits do," she finally says, "is work their faces. Why it's called rabbit tobacco is they out there in that field all the time working their faces."

"And that's what you smoked."

But all she says to that is, "Hunh!"

MR. JEFFERSON'S NUT TREE

Pecan

■

T*hop!*

Gunshot! I jump.

Thop!

Oh, it's our neighbor Al wielding his .410. He's at war again with the squirrels. "Damn critters," he says, fire in his eye and vengeance in his voice. "They're up there in my pea-can tree just a-wreckin' everything as usual. But not this year they don't."

They'll take what they want, though. They always do. Yet, there are enough of what Al calls "pea-cans" to go around. He'll get his share, plus some to give away to anyone who wants to bother with a puny, hard-to-crack variety. Protecting the nuts seems more an excuse to pot away at the squirrels than to save a crop. Anyhow, the squirrels are more edible than the nuts produced by this particular tree that grows in Al's yard just eighty feet in from the banks of the river Neuse.

But why concentrate on one sorry example when the pecan in general merits a place in the highest ranks for the ways in which it gives delight to humankind: graceful sturdiness and dense green shade, wood useful in decorative cabinetry or giving a hickory zest to smoked meat, and—best of all—nuts so good that it's not possible to stop eating only two or three. It was one of Thomas Jefferson's favorite trees, and deservedly so. A man as passionate about plants as he was about life, liberty, and the pursuit of happiness, he almost invariably included pecans in the annual garden plans for Monticello, his hilltop home in Virginia's piedmont, and he wrote of them often and fondly in his letters. Spelling the name in many ways—"paccan," "pacan," "peccan"—and sometimes referring instead to the "Illinois nut," he requested supplies of seed every year, from the 1780s into the early 1800s, and faithfully recorded the dates of planting. At every opportunity, Jefferson promoted this native-American tree both at home and abroad. And from his pen issued the first Latin description of the tree, which earlier botanists had somehow neglected to dignify with the proper scientific language. In his *Notes on the State of Virginia,* written in 1781 to appease the curiosity of a member of the French legation in Philadelphia, Jefferson had this to say:

> Paccan, or Illinois nut. Not described by Linneaus, Miller, or Clayton. Were I to venture to describe this, speaking of the fruit from memory, and the leaf from plants of two years growth, I should specify it as the *Juglans alba, foliolis lanceolatis, acuminatis, serratis, tomentosis, fructu minore, ovato, compresso, vix insculpto, dulci, putamine tenerrimo.* It grows on the Illinois, Wabash, Ohio, and Mississippi. . . .

The Latin translates into: white Walnut, with toothed, hairy, lanceolate leaves gradually tapering to a point and a comparatively small seed that is ovate, compressed, very lightly grooved, sweet, and thin husked."

The pecan was also fancied by some of Jefferson's contemporaries, among them George Washington, whose journal for May 1786 records the planting of a row of "Illinois nuts" in his botanical garden at Mount Vernon. Rumor, currently spread by the otherwise reputable Audubon *Field Guide to North American Trees: Eastern Region,* has it that Jefferson supplied these nuts and that several of the resulting trees survive to this day. Would that such a delightful tale could be believed, but it's just not so. To begin with, Jefferson was in France in May 1786 and had been there for two years. Also, the horticulturist now presiding at Mount Vernon reports that the two great pecans found on the estate are indeed venerable trees—but not quite venerable enough: botanists have reliably estimated that they were planted in the mid-1800s and are thus about 150 years old. The horticulturist commends these "happy trees," which have grown so tall, and are so unlike the shorter, stubbier varieties found in plantations and back yards, that visitors can hardly believe such towering giants are truly pecans. But, in his opinion, Jefferson cannot be entirely dismissed as a provider of nuts to Mount Vernon; there's a live possibility that he donated a quantity that was delivered from Philadelphia in 1794.

Jefferson's letters and garden records are silent, however, on this particular matter and on one other. Although he has bequeathed much information to the present day, although his delightfully unsettled spellings of the tree's name are amply preserved, there is no record of how he said the words that he inscribed on paper with his feather pen.

Pea-can vs. p'*cahn*—a longstanding dispute. One may as well ask, To-*mah*-to or to-*may*-to (which Jefferson also planted at Monticello and spelled "Tomata"). In both cases, the answer varies according to the region of the speaker. When it comes to Beefsteaks and Better Boys, people from England, New England, and the South are more likely to opt for the supposedly more cultivated short a, while the rest of us, including me, just bleat the word. With nuts, though, it's southern country folk who may think that the p'*cahn*-sayers are putting on airs. To complicate matters further, there's a third school that combines the other two pronunciations into p'*can*. The old-time Algonquian-speaking Indians—Crees, Ojibwas, Abnakis—who gave us the word in the first place, might well have recognized any current pronunciation of the nut tree's name; it was their term *pakan*, or *pagan*, that was picked up in the early 1700s by French settlers in Louisiana and given the spelling *pacane*. (In the middle of that century, New Orleans Creoles, exhibiting the usual French interest in gustatory possibilities, put *pacanes* into the chewy-sweet brown-sugar-and-nutmeat confection known as a praline.) But there's no help here on how "pecan" should sound. And it seems that not even the botanists can provide any guidance on pronunciation. Into the twentieth century, some of them called the species *Hicoria pecan*, while others opted for *Carya pecan*, but members of both camps may have disagreed on the proper handling of the vowels and word accent of that species name. Today, they dodge the issue altogether by calling the tree *C. illinoensis*, the Illinois hickory. *Pea*-can, p'*can*, or p'*cahn*? My preference? The last slips most easily from my tongue, for it's the pronunciation of my bringing up, but I can use the first if the occasion so demands.

Illinoensis—the species name refers to the tree's home range in the Mississippi River valley. And that's where it was first noted by Europeans. In 1541, after the Spanish explorer Hernando de Soto and his party had crossed the great river into what is now Arkansas and had finally slogged through the bottomland sloughs and swamps to higher ground, they came across a fine stand of the trees, which the party's chronicler duly recorded and described, calling them "walnuts" for lack of a more precise name; his note on the nuts' thin shell gives away their true identity. After that, as adventurers and settlers made their way up and down the Mississippi valley, they made awed mention of giant pecan trees, with the very largest found in the eastern part of Texas (a place known then as well as now for claiming the biggest and the best). Texas subsequently adopted the pecan as the state tree. The greatest of these trees could attain a sky-sweeping height of 120 feet and a solid girth of thirty feet. It would take five men to encircle a tree that big with their outstretched arms. And the cool green dusk beneath the summer leaves of such monumental old trees must have greatly rejoiced weary travelers. The nuts were even more impressive than the trees, and more desired: sometimes the giants would be cut down, sometimes whole groves would be felled, so that their crop could be easily harvested.

No matter which twist is put on pronouncing its common name, the pecan ranks right at the top as a true-green, all-American tree. So do the other hickories, some seventeen species altogether. Except for a solitary species found in southern China, the hickories now living originated in the New World and claim its territory, from Canada south into Mexico, from the Atlantic coast into east Texas, as their very own. The word "hickory" is also as American as the genus; it's an English version of *pawcohiccora*, the term used by Indians in the Virginia colony for a drink made by mixing the sweet-tasting nuts—shells as well as meat—with boiling water and pounding the mixture into a thick and oily milk, which was strained and allowed to ferment. It could be drunk down straight, poured like gravy over venison, or used with cornmeal to form cakes.

The scientific names of the hickories and their close relatives are another story. When it comes to the formal designations for these trees, the nomenclators reached (as they often did—and still do) into Latin and Greek. The hickories belong to the Walnut family, the Juglandaceae, the name of which comes from the Latin phrase *Jovis glans*—Jove's acorn—which was used by Pliny the Elder, the Roman naturalist of the first century A.D. Oh, the walnut must have been a strange and marvelous phenomenon in Pliny's eyes—and on Pliny's tongue—so tasty and oily rich that it was truly fit for the king of the gods. Why else would he have taken *glans*, a word precisely denoting the smooth nut of an oak tree, and placed it on a capless nut housed in a hard, wrinkled shell? After all, acorns and their parent oak trees, sacred to Jove, were hardly unfamiliar to him, and he had another perfectly good Latin word, *nux*, to use for "nut." The reason may well be that these god-favored nuts did not grow on trees indigenous to Italy but had to be imported. And "walnut" itself, the common name long used in places farther north, has an etymology that speaks of an exotic origin for the tree and its delectable fruit: the Old English words behind the modern term mean "foreign nut." From antiquity into more modern times, then, walnuts have been a prized item of trade; the East has supplied, the West has gratefully received. But over the centuries, the walnut species, which are variously native to North America and Asia, have become widely naturalized in Europe. Or, it might be said that they've come home, for the fossil record shows that Europe once had walnuts of its own but lost them when glaciers came down from the north like great bulldozers in the last Ice Age. As for *Carya*, the genus of all hickories, that word from classical Greek really and truly means "walnut" and will do in a pinch to designate the hickory, a New World tree that never had a proper Old World name.

Botanically, the pecans (indeed, all the hickories) are monoecious, or "single-housed." Both male and female flowers appear on each and every pecan tree; they are uniformly tiny and pale green but clustered apart, the pollen bearers here and here, the seed producers over there. The latter tuck themselves away almost unnoticeably on twig tips amid the burgeoning leaves. But the male blossoms are not reclusive. Appearing three to a stalk in catkins that may be as long as five inches each, they dangle from the tree like fuzzy, ragtag fringes of chenille. But, though they occur on the same tree, male and female do not open at the same time. In some varieties, the staminate flowers reach maturity first but wither before the pistillate flowers are receptive; in others, the opposite happens, with the female ripe, ready, and gone before the male has begun to bloom. In the wild, this strategy has assured cross-fertilization, one tree with another, since the dawn time of the hickories, and has allowed them access to the greatest possible local gene pool.

But in the domestic realm, humankind has practiced unnatural selection ever since the development of agriculture some ten thousand years ago. Like grains and other fruits, pecan trees have been sorted through, some rejected as not worth bothering with and others chosen for characteristics with special appeal to people—the thinness of the nutshell, for example, and the relative absence of its red lining, which may cling to the nut and give it a bitter taste. Botanists think it likely that the Indians of the Mississippi valley picked and chose among the trees available and that they not only favored those that were on the short, stubby side but purposely trimmed them for ease of harvest. Only in the twentieth century, however, has the growing of pecans achieved dependability; only now can we truly select the trees that will bear nuts of the choicest types.

In Jefferson's day it was catch-as-catch-can. The pecans that he put in at Monticello were a far cry from those planted today. The nuts sent to him yearly were wild nuts gathered from trees that had sprung up wherever they pleased in the woods of Illinois and other locales up and down the Mississippi valley. Thick shells and thin, smooth and rough, shapely and gnarled, with sweet or tasteless meat or maybe worms inside—it's certain that those nuts gave ample demonstration of every virtue possible to a pecan, and every last flaw as well. It's also certain that they possessed virtues enough to keep Jefferson planting them year after year, in his Virginia gardens and orchards and in France as well.

In 1784, Jefferson left Monticello to serve, in conjunction with Benjamin Franklin and John Adams, as minister plenipotentiary to France. He did not return to his hilltop house until 1789. While abroad, he took it upon himself (no doubt with great energy and joy) to acquaint the French with many admirable plants native to North America and to introduce them as widely as possible into the gardens of France. To these ends, he often sent letters home to ask for seeds or specimens of everything from azaleas and mountain laurels to tulip trees, pawpaws, and persimmons. One such letter, written in Paris on January 3, 1786, entrusts his friend Francis Hopkinson of Philadelphia with a special commission:

> it is to procure me two or three hundred Paccan nuts from the Western country. I expect they can always be got at Pittsburg, and am in hopes that by yourself or your friends some attentive person there may be engaged to send them to you. they should come as fresh as possible, and come best I believe in a box of sand. . . . I imagine vessels are always coming from Philadelphia to France. if there be a choice of ports, Havre would be best.

A far-reaching commission. Jefferson's request provides evidence that the pecan had not expanded much beyond its original range in the more than two hundred years since de Soto's chronicler had noted it. The nuts would have been collected somewhere in the Midwest—in southern Illinois, perhaps, along the Wabash or the Mississippi—and brought by boat up the Ohio River for the eventual landing at Pittsburgh. From there, they traveled overland to

Philadelphia, where they were put aboard a ship to cross the Atlantic. A long journey, and when they finally arrived in Paris, Jefferson would have handled each nut separately, checking it for viability above all else. I have no doubt that he was a most effective agent for the pecan's dispersal and naturalization well outside the confines of its native habitat.

But Jefferson had to count on a good eye and his feel for quality to tell him which nuts were better than others not just for sprouting but for producing true to type. One of the problems with wild nuts—at least, as seen by human beings eager for perfection—is that wildings may beget offspring quite unlike themselves; a tree that bears small, knotty nuts may grow from a large, ideally thin-shelled seed. The reason for this turn of events is the pecan's use of cross-pollination as a reproductive strategy: ma's grace—and her crooked nose—pa's strength—and his bow legs—and their offspring may show the best or worst of either side.

Grafting solved the problem. In the same fashion that other fruits and nuts are each enabled to retain their special varietal characteristics—their colors and flavors and scents—a twig from a tree exhibiting desirable traits is implanted in a wild rootstock. But this happy union of scion with stock did not occur in Jefferson's lifetime. The first successful graft was made in 1846 by the slave who served as gardener to the governor of Louisiana. For the next sixty years other horticulturists experimented with grafting techniques in hopes of domesticating the pecan; their efforts, however, did not bear much fruit until well into the second decade of the twentieth century. Now, on the eve of the twenty-first century, an abundance of cultivated varieties—"cultivars," for short—are planted, wherever climate and soil will support them, from Virginia and parts south through Texas and on into California. Hastings, Jackson, Mahan with exceptionally long nuts, Stuart famed for a papershell that can be cracked in the hand—these are a scant few of the named standards. And there are the so-called "Indian" varieties—Kiowa, Pawnee, Cheyenne—that grow on smaller trees and bear relatively thick-shelled nuts. (What's in a name? While it's unquestionably right to honor Native Americans by naming varieties of an all-American tree for some of their nations, it would have been better still to include the Indians who gave us the name "pecan" in the first place: the Crees, Abnakis, and Ojibwas. Perhaps some horticulturist, working now and tomorrow for even more desirable strains of pecan, will correct the omission.)

I visit Al's tree. Though it's not a tall tree, its growth has been vigorous and straight; its crown is shapely, and its trunk just too big to get my arms around. But in the matter of its nuts, the tree seems a throwback, exhibiting features that must have sprung from a rude, wild ancestor rather than from a supposedly well-bred scion. This year, as last year and the year before, it will bear gnarly, undersized nuts, with shells as hard as rocks, and this despite the fact that all the other pecan trees in the neighborhood—one of which had to act as this tree's pollinator—are, every one of them, among the more elegant cultivars. It's September, and these inelegant nuts still cling to Al's tree in their lime green husks. It will take them another month or six weeks to fall.

What falls right now is a fairly stout twig, gliding down on leafy wings. The break is so clean that the twig looks cut from the tree with a tiny saw. In a sense, it was. The larva of a wood-boring beetle—the oak twig pruner, perhaps, or the painted hickory beetle—working neatly beneath the bark has separated the twig from the tree, and all it took was a slight gust of wind off the river to sever the tenuous connection. The beetle-to-be, snugged inside, rides with the twig to earth; instinct has commanded it to wait for spring and metamorphosis within the safety of a wooden house. But it's out of luck, for the weather hasn't begun to close in enough to keep the grass from growing. Al comes along on his lawn tractor and shreds the twig—leaves, wood, tenant, and all. When he spots me, he turns off the engine.

"I saw that," he says, nodding at the slight scattering of debris. "First squirrels, then sawyer beetles—what next? This is what you get, though, with a wild tree."

"Get those problems with a tame tree, too," I say. Then the lightbulb goes on. "Wild tree . . . you're saying that this one's not tame?"

"Not hardly," he says and tells me what happened two decades ago when the tree was newly planted and hadn't yet begun to bear. His son ran over the sapling with a lawn mower. Al tried to rectify the damage with clippers and pruning shears. "Didn't work. I cut away the graft and left the root. And that's what's been growing here the last twenty years—that old root, putting out those damn hard little *pea*-cans and raising a right healthy crop of squirrels."

I am unexpectedly thrilled. Thomas Jefferson would have recognized this tree. He might not have chosen its nuts for eating and certainly not for planting, but he would have known this tree, known it for exactly what it was and is.

PECAN: THE INSEPARABLE ADJUNCT

Certain words are not complete in themselves but must be part of a pair. What is higgledy without piggledy, upside without down, or Jack without Jill? In my view, "pecan" does not stand alone but needs the word "pie." So transformed, the nut reaches its highest potential. The pie is easy to make and suitable for every high and low occasion, from Thanksgiving dinner with all the fixings to plain old neighborhood potluck. The following recipe is my variation on the excellent instructions that appeared fifteen years ago on every label of Karo corn syrup. The nutmeats may be crushed, but I prefer to use them in showy halves.

Pecan Pie

Ingredients

3 eggs, slightly beaten
1 cup dark corn syrup
1 cup sugar
1 1/2 cups pecans
2 tablespoons butter, melted
1 teaspoon vanilla (or 1 tablespoon rum)
1/8 tablespoon salt
9 inch deep-dish pie shell

1. Mix together eggs, corn syrup, sugar, melted butter, vanilla, and salt. Stir in the pecans. Pour into the pie shell.
2. Bake at 400 degrees for 15 minutes. Reduce heat to 350 degrees and bake for 30-35 more minutes. The filling should be slightly less set in the center than around the edge.
3. Cool. Top with whipped cream, if desired.

THE APPLE ANCESTOR

for my grandmother Jannette

"When I am dead,"

she said, often those darkening
days of her ninety-first autumn,

"I'll have no foolishness. Put me
in a plain wood box that opens easily
to worms and bury me in rich red loam

on a sunny slope, well-drained. Don't
burden me with marble but plant me
an apple tree, if you please.
Winesap. That's a good late keeper."

In its long winter, grief

would have so pleased and set in
turned earth the stock and the scion

that there, in rust, the grey sticks
might sleep through funeral dark;
there, lightless, begin to root,

wake the dead eyes, tap the dry womb,
so that equinox come, the lightening
of mourning, spring urging me lifeward,
I would see her greening,

nourished, translated: twigs
quick with leaf, brideclouds slowly
exploding; year on year growing;

branching, fertile, toward noon's pippin
or Eve-bitten moon; and bending,
complete, to the green hill.

And in my old wives' summer the fruits
of her first living three times renewed
might climb and pick full-breasted red
ripeness and bite tart good sense.

Other flesh savors stones.
I plant her these words.

WHERE, OH, WHERE IS PRETTY LITTLE SUSIE?

Pawpaw

■

Where indeed? Sometimes this question, well known to folksingers and Girl Scouts, is asked about Sally, or Liza. Whatever her name, she's always pretty, and the answer always puts her "Way down yonder in the pawpaw patch."

I don't know about Susie and those other girls, but that's where I've put myself on a bright blue day early in September. Pawpaws to the right of me, pawpaws to the left of me, the trees in this patch march in serried rows. The other people here bend and reach, busily picking the ripe specimens (or so they hope; we'll put them to a taste test in two days). I, however, stand bemused in the huge green shade of densely clustered leaves. There's half a pawpaw in my right hand. My mouth is full of pawpaw. And sweet, custardy pawpaw is plastered on my cheeks and chin; pawpaw juice runs down my arm and drips off my elbow. This is the first one I've ever tasted. I'm breathing pawpaw, too, for the air is gently but insistently perfumed by the maturing fruit. Butterflies—black swallowtails, pearly eyes, red-spotted purples—flit everywhere.

Both butterflies and people have been summoned here by sweetness. And all of us have gathered to celebrate the ripening of *Asimina triloba* by eating as much of it as we possibly can. The occasion could not be more festive. Nor more festively earnest: the people have an additional agenda. They're laying plots that may affect the place in which pretty little Susie has long been known to hang out. But of that, more later.

Pawpaw, what a silly name! It pats itself with quick strokes. It puffs out the cheeks. It sounds like blithering or scat singing. It makes little, breathy explosions, pop-pop. It babbles like baby talk. And the fruit looks somewhat silly, too—a bulbous, kidney-shaped handful that's green with bruise-purple streaks and patches on the outside, creamy white to orange on the inside, and studded liberally with hard, dark brown seeds the size of fordhook lima beans. *Thhp!* The seeds are almost as easy to spit as their smaller, lighter watermelon counterparts. But it's not really fair to characterize the tree by either its risible name or the homely appearance of one discrete part. The pawpaw is a plant of many and amazing uses.

I encountered my first member of the species in my neighbor Mo's yard, a place in which I often meet odd sorts of green growing things, from elephant garlic to sassafras to warty bark (a local name for the hackberry). Nor is it just a single pawpaw tree that grows out back by Mo's shed but eleven of them. The grove is doubly shaded, for the pawpaws cast their own shadows over the ground, and they themselves grow to perhaps twelve feet beneath a dark green canopy of far taller oaks and maples. The tree trunks are slender—I can almost circle the stoutest of them with thumb and middle finger—but the smooth-edged leaves, though narrow, are enormously long. From stem end to pointed tip, a single leaf can span the distance from my wrist to my elbow.

Books warn that the leaves emit a highly disagreeable odor if they're bruised. The only proper response to such a warning is, of course, to pluck a leaf and crumple it between fingers and thumb. The scent is lightly acrid and lemony, overlaid at first with a hint of something fetid, like the smell of muck at low tide, but the overlay soon dissipates, leaving an aroma that I find pleasantly astringent. It's the pawpaw's beautiful flowers that wrinkle the nose and purse the mouth with a light but appalling scent.

The flowers bloom in early spring, before the leaves begin to emerge. They dangle from the slim branches like bells, and each bell has three sepals in its flower cup, three petals around its flared lip, which spreads nearly two inches across at the peak of blooming. These attractive tripartite arrangements are responsible for the second part of the tree's binomial, *Asimina triloba*—three-lobed Asimina.

Mark Catesby (1682–1749), an Englishman who devoted considerable study to the natural history of the southeastern American colonies, deserves credit for coming up with both barrels of the pawpaw's binomial, which was adopted without change by Linnaeus, the father of modern taxonomy. But nobody now knows where Catesby might have found *Asimina*. I've heard that it means "sleeve-shaped fruit," heard also that it's a Greco-Latin term for cinnamon, and heard that the French bestowed this name on the plant. I came across the first guess in *The Tree Book*, written by Julia Rogers in 1905; it's a thin-air notion without discernible ancestry. The second guess also lacks substance (Catesby knew his Latin), but the third may have merit (he spoke French). It could well be that early French explorers, not knowing what else to call a plant completely unfamiliar to the Old World, picked up its local name, which Catesby later latinized. Like persimmon, catalpa, tupelo, sassafras, and other American tree names, Asimina is—I'd bet on it—an adaptation of an Indian word. As for *pawpaw*, responsibility for that name is easily assigned; it rests with the sixteenth-century Spanish adventurers, who mistook the pawpaw for a somewhat similar New World fruit that also has greenish skin and orangy flesh, the papaya. Over time, language played its usual tricks, transforming *papaya* into *pawpaw*. (And how did the Spaniards come across the word *papaya*? The same way in which the French found the name ancestral to Asimina—from Native informants.)

That nose-wrinkling scent: its foul nature is announced by the flower's color—red to purply brown, the color of spoiling meat. Not unexpectedly, the smell matches. But the pawpaw, unable to consider the effect of its perfume on the human nose, is simply fulfilling a built-in command to attract the carrion-loving flies and beetles that act as pollinators.

Mo says that his trees produce abundant blossoms but a scant set of fruit, and that's been a problem with his pawpaw patch ever since the first four trees that he got from his good friend Dagger Caudill were planted over by the shed. Dagger Caudill dug up the four in his home state, Kentucky, and brought them south to the Carolina coast because Mo is the kind of man who would appreciate a gift from the wild. The other seven trees in the patch have sprouted up from runners sent out by the original quartet. They bloom just as merrily as their progenitors, and they set just as little fruit. Last year Mo tied a long piece of string on a lower limb to identify the one tree that gave him any pawpaws at all, including two that actually achieved ripeness. This year we see fruit on four trees. Most of it grows in clusters of two on a single stem, and each twosome spreads outward in opposite directions like a pair of wings. One cluster, though, boasts five green fruits, the smallest the size of a pecan, the biggest closer to a duck's egg. By the end of July, however, every last fruit, still green and hard, has dropped to the ground. Insect damage? Six weeks of little rain? An aversion to the salty air on this Carolina coast? We speculate but arrive at no answers. My first taste of pawpaw must willy-nilly be postponed to a later time.

A most peculiar tree, the pawpaw—and not just in respect to its silly name and a flower possessing an odor that contradicts its beautiful appearance. Granted, these peculiarities are not inherent but rest in the judgments of an observer. But the pawpaw is an odd tree in its own right, leading a far less restricted life than most of its family.

The pawpaw belongs first of all to the Magnoliales, perhaps the most ancient order of angiosperms, plants that form seed within a protective ovary. (The order's designation comes from the name of one Pierre Magnol, a French botanist, who died in 1715.) The pawpaw, a tree native to North America, is more particularly a member of the Annonaceae, the Custard Apple family of some 130 genera and more than two thousand species, which is found almost exclusively in warm or downright tropical habitats in both the Old World and the New. The family name sprang from the classically educated imagination of an old-time nomenclator, very possibly Mark Catesby, whose designation *Annona glabra* was selected by Linnaeus as the binomial for the pond apple, a pawpaw cousin found in Florida. And of all the Annonaceae, the pawpaw is the rule breaker. Unlike every other plant in this large family, it doesn't just tolerate but thrives under conditions that include snow and ice. An adaptable tree, it makes itself at home in many climates, from the generally temperate weather on the mid-Atlantic seaboard to the hot Julys and cold Januarys of eastern Kansas, from subtropical steam in Georgia and Louisiana to the frigid winters of Michigan and southern Ontario. Nor does the pawpaw exhibit much discrimination between high and low places but grows at many elevations. It can be found in the Appalachians and on the Texas coast.

Word of the pawpaw may have first reached Europe through the Spanish explorer Hernando de Soto. According to the chronicles of his expedition, which swashed and buckled its way through the South on a great treasure hunt in 1540 and the following year, he and his men met Native inhabitants who were said to have cultivated the tree for its fruit. Its first mention in English may be that in John Lawson's 1709 *Natural History of Carolina*:

> The Papau is not a large Tree. I think, I never saw one a Foot through; but has the broadest Leaf of any Tree in the Woods, and bears an Apple about the Bigness of a Hen's Egg, yellow, soft, and as sweet as any thing can well be. They [the Indians] make rare Puddings of this Fruit. The Apple contains a large Stone.

Specimen trees were sent to England in 1746 by the botanist John Bartram. Patrick Gass's firsthand account of the Lewis and Clark expedition, published in 1807, a full seven years before the official record was issued, spoke of getting "a great many papaws, a fruit in great abundance on the Missouri from the river Platte to its mouth." And in *Domestic Manners of the Americans,* her 1839 compendium of our New World crotchets and foibles, that tireless observer Mrs. Frances Trollope remarked that "near New Orleans the undergrowth of palmetto and pawpaw is highly beautiful." In a painting from the early 1820s, John James Audubon also acknowledged the pawpaw by placing two yellow-billed cuckoos on its branches; there, beside a cluster of seven fruits, amid a mass of leaves (chewed mightily by insects), the birds flourish their wings and tails.

Papaw—the spelling used by Lewis and Clark's Mr. Gass is the one preferred by dictionaries and encyclopedias. Look up *pawpaw*: you'll be bounced forthwith to an entry headed by the shorter version. I don't know why, but it hardly matters; a pawpaw by any other spelling is still a pawpaw. And it's still a pawpaw by any other name, of which it can lay claim to a fair variety—Indian banana, Michigan banana, dog-apple. I note that these names zero in on the succulent sweetness of other fruits, as if the flavor of a pawpaw is so beyond the range of human words that it cannot be truly described. And the name of one of these other fruits pops up in the family's common name, Custard Apple, which speaks of the creamy texture and sugary taste of all the Annonaceae.

What is their sweetness like? What is the allure of that taste I won't taste till I go at last to the pawpaw patch? In *The Tree Book*, Miss Julia Rogers offers a caution (which reflects a prejudice not atypical for 1905):

> The pawpaw's soft pulp, in its green banana-like envelope, is delighted in by the Negro of the South. It is sold in the markets, but is too sweet and soft to be really enjoyed by more fastidious people. One must get used to the pungent pawpaw taste, and then only the yellow-fleshed fruits are fit to eat. These are improved by hanging on the tree until they get a sharp bite of frost.

But how am I to believe a woman who, in the next breath, delivers the dubious rumor about *Asimina* meaning "sleeve-shaped fruit"? I visit Mo's barren trees, acquire dry tidbits of botanical and historical information, and yearn for a sample of earthly ambrosia.

The way to the pawpaw patch is surprisingly easy. A newspaper features a story about a man named R. Neal Peterson, who is so happily obsessed by *A. triloba* that he has become the pawpaw's counterpart to Johnny Appleseed. The article is illustrated by his portrait: long, thin face with merry eyes, broad and slightly off-center smile, and a full but well-trimmed mustache, all topped by a hat so shapeless that anyone else would have thrown it out. He looks like the king of the road, but he's really an agricultural economist who works for the U.S. Department of Agriculture and vigorously promotes the cause of the pawpaw in his spare time. I write to him in care of the agricultural research station mentioned in the article. There, and in several other locations in the mid-Atlantic region, he has planted not just a patch but a whole orchard of pawpaws. The next thing I know, I'm a member of the PawPaw Foundation, one of the fruits of his obsession, and I have an invitation to attend the world's first-ever pawpaw conference.

The conference takes place Friday through Sunday on the second weekend in September. And the gods that smile on pawpaws —and on the people who fancy them—fill the weekend with sunshine and fragrance. The nominal location is Keedysville, a hardly noticeable crossroads six miles from the Civil War battlefield at Antietam; the actual location, the gently rolling four-hundred-plus acres of the University of Maryland's Western Maryland Research and Education Station. On the horizon the Appalachians rise like blue mist. Forty-three people show up for the occasion. Some are botanists with university affiliations and some, professional nurserymen or seed-exchangers or writers on horticulture (not gardening, mind you, but horticulture). Name tags identify two others as "Oldtime Pawpaw Fancier" and "Serious Pawpaw Eater"; both turn out to be reliable sources of factual information. (Neal's name tag reads "Mahatma Pawpaw.") Our number also includes a few camp followers—one spouse, one mother, and one eternally curious amateur, me. We come from near and far—Georgia, Arkansas, Michigan, Oregon, Ontario, the Crimea. One woman represents the other side of the world; she's a member of the Institute of Forestry in Beijing, China. All of us behave as if we're at a party, and, clearly, we're all a little bit crazy. The conference is swept by galvanic waves of energy and excitement that one partygoer, Colleen Anderson, fondly characterizes as "spontaneous outbursts of silliness." But what other than silliness can be expected of people who gather to celebrate the pawpaw?

And not just celebrate it but conspire to promote it to the deprived, the ignorant, and the unbelieving. Of course, the conference has an agenda, not the hidden kind but one declaring itself without the least hesitation. By the time we hit the halfway point, almost all of us are running around in bright gold extra-large T-shirts that shout the conference slogan in bold brown letters:

I'm
pro-pawpaw
and I vote.

And by that time, I've learned all sorts of things that aren't on the official program. When I'm back on the Carolina coast, I'll be able to tell my neighbor Mo not only about pawpaw reproduction and pawpaw cousins in the Custard Apple family but also about the pawpaw and the pepper, the apparent predilection of crawfish for pawpaws, the probability of pawpaw-scented perfume, and the procedure by which a whistle can be fashioned from a pawpaw twig. I'll tell him, too, that under certain conditions the pawpaw becomes positively obscene.

Obscene? So says one pawpaw partisan, a botanist from the University of Georgia. Right after signing in, we've made an excursion to the orchard that Neal Peterson planted in 1984. Here the novices among us are supposed to become acquainted with evaluating green-skinned fruit for its readiness to pick. None of it looks really ripe. But —handing me the pawpaw that soon has me sucking and slurping while other conferees are diligently filling paper bags with fruit—the botanist grins and elaborates. "Oh yes. When you get 'em closed up in a car, that's when they're obscene, so fragrant, aromatic, and . . . *attractive.*"

The orchard itself scents the air for a good half mile in every direction. Closed up, the fragrance could indeed have knock-out potential. Light but heady, it accounts for the presence of the perfume chemist. These days perfumers search greedily for what they call "food notes"—the odors of apples, apricots, peaches, and other fruits—to enhance their concoctions. The chemist is here gathering *A. triloba* so that, a mere three days hence, the chemical compounds producing its distinctive smell can be analyzed, re-created, and confined in bottles. And when the bottles are unstoppered and the genie is released—pow!—women will smell as magically, seductively, obscenely sweet as a pawpaw at the peak of ripeness.

I walk with Colleen from the orchard back to the station's conference facilities, the site for our scheduled talks and workshops. The tall, dry grass beside the lane is ornamented by Queen Anne's lace, and catbirds mew softly, querulously, in the hedgerows. As we mosey along, Colleen hints at the adventures of the pepper and the pawpaw. I ask for more than a hint. A board member of the PawPaw Foundation, editor of its occasional newsletter, and designer of the loud gold T-shirt, she's also a singer and a writer of lyrics in the folk tradition. And she proceeds to give me a rollicking sample. The tune resembles that of "Sweet Betsy from Pike."

> The pawpaw and the pepper were riding in the car.
> Said the pepper to the pawpaw, Are we going far?
> Said the pawpaw to the pepper, No, we're driving into town
> To do a little business and take a look around.

All right, but what happens next? She promises to sing the whole story when there's an opening in the program.

The next day's workshops and talks go a long way toward appeasing my curiosity. The problem is that the sessions are too short for total satisfaction. The speed of presentations is zip-zip-zip. These impassioned partisans of a peculiar fruit could give long-winded speakers great lessons in brevity and keeping an audience fully alert.

Neal shows slides of water tanks bearing the word that has

brought the partisans together: Paw Paw, West Virginia, and Paw Paw, Illinois; this town name, which is also displayed on high in several other states, attests to the esteem in which the fruit was held by pioneers pushing westward. And such esteem was warranted, he adds, though no one could have been aware back then that the pawpaw wins out handily over the peach, the apple, and the grape in just about every nutritional category. It's low in moisture, relatively high in protein, and high indeed in carbohydrates and food energy. An excellent source of vitamin C, it also contains A, thiamine, riboflavin, and niacin. In the mineral department, it abounds in potassium, calcium, phosphorus, and iron. And when it comes to the essential amino acids, the pawpaw ranks right up there with the perfect egg white. A fruit for all appetites, a wonder fruit indeed—except for the blunt fact that although apples, peaches, and grapes lose to the pawpaw in the comparative nutrition stakes, they win the prize for availability every time.

In another presentation, I meet the daintier Florida members of the pawpaw's genus, such as *A. pygmaea* with four-petaled white flowers and *A. obovata,* also called Florida dogwood, the white blossoms of which smell uncharacteristically sweet. Not one of the seven subtropical species grows so grand that it might be called anything other than a shrub. One of these, *A. tetramera,* is so rare that it might be considered endangered. But like their close cousin, the more robust pawpaw, all bear edible fruit. And they share a reproductive problem with the pawpaw: an extravagance of flowers, a niggardly set of fruit. In wild pawpaws, for example, only .5 percent—half of a percent—of the many blossoms ever manage to produce fruit.

Something seems to be amiss with strategies for pollination. Part of the difficulty stems from the fact that the genus *Asimina* is programmed for protogyny, a term that translates roughly as "ladies first." The female parts of each flower, the pistils, mature earlier than its male parts, or stamens, and not only do they ripen first, they also close their doors to all further business before the stamens of the same flower begin to shed their pollen. Protogyny aims, of course, at avoidance of the botanical equivalent of incest, as does its opposite number, protandry, or males first. Nor is protogyny unique to the pawpaw. It is practiced by many plants, including the apple and the pear; it's the reason that orchardists and backyard fruit fanciers put in two varieties of pear, for example, so that in the next window of opportunity, the pollen from one tree is available to fertilize the other's female receptors. Two compatible pear trees assure a bumper crop. Why, then, should pawpaws fail?

Pawpaws hardly grow in isolation, one here, one far over there. They're sociable trees that flourish in one another's company; they're given to growing in the famous patches where Susie hangs out. But something's lacking: despite the flowers' spoiled-meat color and their carrion scent, they aren't truly efficient at attracting the flies and beetles that serve as pollinating agents. One of the pawpaw partisans suggests hanging roadkill in the branches. The conferees roar but soon return to the serious business of how one coaxes a bumper crop from reluctant pawpaws. The answer is hand pollination. How will my friend Mo respond to that? It means keeping a vigilant eye on the ripeness of pistils and stamens and effecting pollen transfer, one blossom at a time, with a fine-haired brush.

During a short break, Colleen teases me with another verse of her song:

> The pawpaw pulled the pickup truck into the drive-in bank.
> He took out all his money and he told the teller thanks.
> Then the pawpaw and the pepper, they went into a bar.
> They had a couple quick ones, and then a couple more.

What next? "Later," she says and ducks into the conference building's kitchen to help the lunch crew assemble sandwich-makings.

We hear two more lightning presentations before lunch. A research botanist outlines efforts to develop the pawpaw as an alternative crop for orchardists and for farmers looking to grow something other than tobacco in this new era of decreasing allotments and increasing scorn. And a plant propagator says that pawpaw is the most requested item year in, year out in the seed exchange that he manages for the nursery at Michigan State University. Recipients are instructed to "stratify" the seeds—that is, to use cold to break their dormancy. Seeds benefit greatly by spending sixty days or so in a refrigerator.

Lunch includes an edible display of the pawpaw's tropics-loving relatives in the genus *Annona*. They have been brought for our wonder and delectation by a nurseryman and self-taught hybridizer who first met the Annonaceae in the Amazon basin of Brazil, where his father served as a missionary. What an odd bunch these Annonas are! Green, white, gold, brown, black—they come in assorted colors and combinations thereof. And they bear intriguing, sometimes exotic names—pond apple, mountain soursop, cherimoya, atemoya. Some skins are dull; others shine. Some have large scales resembling those of a snapping turtle's domed shell; some sport prickles; one is as knobby as a hand grenade, while another brandishes sharp, claw-like hooks. Inside, the creamy flesh is variously white or peach or gold. The flavors range from slightly acidic through bland to saccharine sweet. Everyone slurps. Everyone wears pulp on lips, nose, cheeks, and chin.

When I go into the kitchen for a wet towel to wipe my face, Colleen is there. We've left the pawpaw and the pepper drinking quick ones in the bar. What then? She yields to pestering and starts to sing.

> They played a game of pinball and they played
> a game of pool.
> But the pawpaw beat the pepper and the pepper
> lost his cool.
> He punched out the pawpaw and gave him quite
> a bruise.
> Then he fell down right beside him and they

both slept off the booze.
They woke up in the morning, side by side,
feeling mean
With a double-decker headache and a taste
for caffeine.
Said the pepper to the pawpaw, I'm a low
and sorry pickle
But I sure could use some coffee and I haven't
got a nickel.

So the pawpaw took the pepper to . . .

She stops singing. "Later," she says and dashes off for the beginning of the afternoon program.

The afternoon sessions and conversations, along with another trek to the orchard, provide all sorts of pawpaw lore. It goes a long way toward appeasing my never-ending pack-rat urge to store away information, no matter how grand or trivial. And here are some of the bright, shiny tidbits I've tucked away.

- This from the Serious Pawpaw Eater: To make a pawpaw whistle, cut a section, half an inch in diameter and about three inches long, from a slender branch. Pound the bark with a knife to loosen it from the xylem (that's what the SPE, not only a dedicated consumer but a technical purist, calls the plant's woody tissue). Slip off the bark. Cut a notch on one side about an inch from one end. Blow.
- Another contribution from the SPE: The Indians used pawpaws as a source of fiber with which to make baskets and fish nets. They'd strip off bark in the spring when the sap is rising and the bark still flexible enough to weave or bend into the pattern desired.
- This from a biologist teaching at King's College in

Tennessee: Many terrestrial animals relish pawpaws—foxes, squirrels, opossums, and the like. Aquatic creatures may also find them delectable. Evidence indicates that crawfish drag pawpaws into their underwater burrows. The biologist has never caught them in the act, but seeds are often found at the burrow entrances, and nearby fruit bears pincer marks.

- This from a board member of the PawPaw Foundation: Ridiculous, the statement in the *Britannica* and other encyclopedias that there are two kinds of pawpaw, one edible, the other not (though ripe may certainly be differentiated from unripe). Pawpaw is pawpaw is pawpaw, though some trees bear fruit tastier than that of others.
- And this from a passerby who wandered into the orchard while visiting the Research Station for other reasons: "Well, I'll be damned. So that's what those things are. Friend of mine throws 'em for his dog to fetch."

It surprises me, however, to find that no one here knows the origin of the family name, Annonaceae. No one except me, that is. But then, no other conferee has spent much, if any, time prowling around in the ancient Mediterranean past. The name pays tribute to the Roman goddess Annona, who presided over the year's agricultural produce and took her own name from *annus*, the Latin word for "year." No Annonaceae grow in Italy, but the classically educated taxonomist who chose that designation may have thought that fruit with a divine taste should be honored with a divine name.

What *is* its taste? Since eating my first pawpaw out in the orchard, I've reached for and rejected a hundred descriptions. The pungency ascribed to the fruit by the fastidious Julia Rogers does not exist (and, despite what she says, the fruit doesn't hang on the tree long enough to be brought to its peak by "a sharp bite of frost"). It tastes uniquely like itself, of course. And it seems unfair—demeaning, almost—that I must resort to other fruit in an attempt to suggest the pawpaw's unparalleled flavor. But this is the closest I can come: the creamy smoothness of banana, enlivened by a light but definite hint of pineapple, a dollop of clover honey, and a dash of vanilla.

It's after supper that the pawpaw's most startling possibilities are presented. With its relatives, it's truly an all-purpose plant, providing food, fiber, fragrance, whistles, and toys for dogs to retrieve. Sometimes it triggers allergies. It can also heal. But first, before we dine, Colleen treats everyone to the full version of her song. And here's what happens when the pawpaw and the pepper wake up in the morning with thumping heads and a craving for caffeine:

So the pawpaw took the pepper to a diner
downtown.
They ordered eggs over easy and a side of
hash browns.
They drank a pot of coffee and they smoked
a cigarette,
And they talked about religion and they talked
about the Mets.

The pawpaw put a penny on the table for a tip.
Then they got back in the pickup and they took
off at a clip.
Said the pepper to the pawpaw, Do you think
you have the nerve
To take her up to eighty on Dead Man's Curve?

The pawpaw and the pepper, they had an awful
wreck.
It squashed the pawpaw flat and it broke the
pepper's neck.
And so my story's ended, with only this to tell:
The pawpaw went to heaven and the pepper
went to hell.

Otherworldly destinations, and, as it happens, the pawpaw may be found these days in another wonderland, the realm of pharmacognosy. The term denotes the science that investigates medicinal substances of plant and animal origin; it deals with the chemical makeup, production, and uses of such natural products. As it also happens, a pharmocognosist from Purdue University has spent the last fifteen years on a process he calls "grind and find." More precisely, he has devoted himself to study of the chemical compounds found in the Annonaceae, with particular emphasis on pawpaws and their *Annona* cousins. After supper, he tells us about his work—and how his interest was explosively aroused: way back in Michigan, when he was three years old, his dad brought pawpaws home. Small boy ate a bunch of them and promptly vomited.

Though only a tiny percentage of pawpaw eaters succumb, the fruit's emetic effect has long been known. Nor are the pawpaw's insecticidal activities any secret to its partisans. But as someone new to the scene, I'm amazed at the pharmacognosist's list of pests repelled by *A. triloba*, if not killed outright: mosquito and blowfly larvae, melon aphids, two-spotted spider mites, cabbage loopers, European corn borers, Mexican bean beetles, and others, many others, including nematodes. The pharmacognosist also mentions a rumor—a bit of folklore, he says—that one Far Eastern species in the *Annona* genus is fatal to head lice if mixed with coconut oil and applied to the hair. It's not folklore, however, but a certainty that the beavers on his farm neither eat any part of a pawpaw nor cut the trees down.

The compounds bringing grief to both pests and stomachs are most heavily concentrated in twigs, unripe fruit, root wood, and seeds. This fact suggests that a wise pawpaw eater will avoid chomping into immature specimens. And the seeds should always be given the *thbp* treatment. As for borers and bean beetles, tests show that insecticides made of pawpaw compounds have consistently proved more effective than either rotenone or the pyrethrins, two widely used natural preparations. Because pawpaws grow easily and well, extracting and marketing their pesticidal compounds would be an economically feasible proposition. In the near future, environmentally friendly pawpaw powder may sit on garden-shop shelves beside the rotenone and bacillus thuringiensis.

The pawpaw's obviously toxic properties have led the pharmacognosist to all-out sessions of grind-and-find. His research has centered largely on what he calls "chemotherapeutically active annonaceous acetogenins." In plainer terms, he's looking at the Annonaceae for compounds that destroy cancer cells. Enough progress has been made so that research has moved up through several of the federally permitted stages, from initial assessments of toxicity to using the compounds on laboratory-grown cultures of tumor cells and on to the current administration of compounds to live animals with tumors, in this case mice. The testing efforts have, not unexpectedly, been interrupted and delayed by some of the more obtuse regulations concocted by the bureaucrats. Illustrating this problem, the pharmacognosist speaks gently and with more patience than I could ever muster: "We're supposed to spend twenty million dollars on tests to prove that the environment won't be harmed by a material that comes from the environment." But given the laudable objective, his gentleness and patience seem warranted. So far, not one but several compounds have shown great promise, particularly for cancers of the lung, colon, and breast. Some of these compounds also attack cancer cells that have become resistant to conventional drugs. Is another botanical wonder drug in the making? Another cancer killer as effective as Taxol, which is extracted from the Pacific yew? The pharmacognosist thinks that the probability is high but cautions that years will pass before the results are in and any drug is approved for human use.

This final evening of the conference holds a valedictory note. It's invested, too, with a faint but perceptible sadness. The first-ever international pawpaw conference may also be the last. With this occasion, Neal's years of work have reached fruition; his pawpaw gospel has attracted converts who can go out and spread the word from Georgia to Oregon and across the world to Russia and China.

These knowledgeable, skilled paw-partisans have already taken on such projects as collecting "germplasm," as botanists refer to genetic materials; defining the best growing conditions; grafting and hybridizing; and turning pawpaws into a fruit as commercially attractive as apples and grapes.

Which leads me back to those paper bags that were filled with freshly picked pawpaws on the first afternoon. Pears, tomatoes, and many other fruits that ripen slowly can be harvested, shipped, and put on the produce counter before they're fully ready to eat. Oranges and grapefruit are picked at maturity and, with proper storage, keep their flavor and texture for months afterward. But the useful life of a pawpaw is short indeed. If September's crop is correctly judged, two days after a fruit is taken from the tree, it should reach its sweetest, creamiest, most obscenely fragrant point of ripeness. (Leave deadfalls on the ground; they're far past their peak.) But two short days after that, a fruit kept at room temperature is no longer palatable. Refrigeration at a relatively high 40–45° may, however, delay the downslide for as long as a week. And after a week? Aged bananas with dark brown skins may be mashed and made into delicious bread; apples that have been on hand for months may be trimmed and turned into sauce. Not so the pawpaw; when it's gone, it's gone—past any hope of resurrection. One of the last acts of the conferees is, yea or nay, to check the contents of those paper bags. The seeds of the hardiest, most flavorful specimens will be saved for future generations.

Where, oh, where is pretty little Susie? She's still down yonder in the pawpaw patch looking for fruit at its optimum moment. But if Neal Peterson and his disciples have their way, the question will get some new answers. At first, we'll hear that Susie's been seen way down yonder at the roadside produce stand. Later, if all goes as planned, we won't have to ask where she is. My neighbor Mo and I shall be right there beside her in the supermarket's produce department.

FROM THE PAWPAW PATCH TO LITTLE SUSIE'S KITCHEN

The second verse of the folksong tells what little Susie's doing after she gets to the pawpaw patch: "Picking up pawpaws, putting 'em in her pockets." I'm sure, though, that she pauses to eat some on the spot. As for the rest, it's reasonable to think that she's a connoisseur of pawpaw cuisine, that's she's always toted them home by the pocketful not just for eating as is but for inclusion in other dishes.

Hard-core fanciers, like Serious Pawpaw Eater, do prefer their pawpaws raw. (It may be that they don't want to waste one minute in transferring that scrumptious flavor from tree to tongue.) They cut the ripe fruit in two (side to side or top to bottom, doesn't matter) and scoop out the pulp with a spoon. Lacking knife and spoon, they've been known to peel off some skin and squeeze out the pulp as if it were icing in a pastry bag.

Pawpaws may be combined with other ingredients to prepare an array of sweet treats. They do not take kindly, however, to a lot of heat, for heat destroys the volatile compounds that create the fruit's distinctive flavor. Use pawpaw pulp in recipes that call for little or no cooking—ice cream, sorbet or sherbet, chiffon pie, zabaglione. For the last, which is especially delicious, a recipe is given below.

Despite the caution to avoid heat, it's a safe bet that Susie has made pawpaw bread. Any favorite recipe for banana bread will do. Simply use an equal quantity of pawpaw pulp in place of the bananas. Nuts—walnuts, hazelnuts, pecans—may be added to the dough.

Pawpaw Zabaglione

This concoction is the creation of Michael Luksa, head chef at the Yellow Brick Bank Restaurant, Shepherdstown, West Virginia. It is the elegant result of a challenge by Neal Peterson that the chef, a man who knows his sorbets from his sherbets, present dishes appropriate to pawpaw people at the banquet held on the conference's first evening.

INGREDIENTS

6 egg yolks
1/2 cup sugar
1 cup whipping cream
1 cup passion fruit liqueur
1–2 pawpaws, enough to make 1 cup puree

1. Remove the pawpaw skin and seeds before pureeing.
2. Combine egg yolks and sugar in the top portion of a simmering double boiler.
3. Heat the liqueur until warm (but not hot, lest it ignite).
4. Add the warmed liqueur to the egg yolk and sugar mixture. Cook, stirring constantly, until thickened.
5. Allow mixture to cool and fold in the pawpaw puree.
6. Whip the cream into stiff peaks and fold it into the mixture.
7. Serve chilled.

Yield: 6 portions

SUDS

The Yuccas

■

Silkgrass, beargrass, Adam's thread-and-needle, mound lily, soaproot, Spanish dagger and Spanish bayonet, Joshua tree—the common names given to various yuccas are almost as many as the uses to which these plants are put.

It's the soap made from yucca that has intrigued me for almost as long as I've been alive. Once upon a time, in the days before World War II, my mother read me a children's book about traditional life in a Hopi village—food, dress, home and hearth, planting corn, kachinas and dancing, the pueblo's sacred mountain. And there was a picture of a woman washing her long black hair. The text said that she'd made her shampoo from the root of a yucca, though it didn't say how. The suds looked creamy and mild, better by far than the eye-stinging stuff (and subsequent vinegar rinses) that my mother used for washing my wispy blond hair. But we didn't seem to have any yuccas on hand. If we did, my mother, who knew just about everything, would have known how to gather them and make that wonderful shampoo.

For years, the desire to test yucca soap lingered, but in a do-nothing fashion. I had become a mother, using Johnson's Baby Shampoo for my own child, before I exerted myself enough to gain information about the soap-extraction process. And it was just that—theoretical how-to information, not hands-on practice.

At that time, two yuccas grew in the weedy garden that came with our rented house near the shores of Lake Erie in Ohio. Long stems like stout reddish stalks of asparagus shot up from both plants and budded out and bloomed with dozens of large, densely clustered creamy white bells. The plants stirred long dormant curiosity, but it wasn't until the *Encyclopaedia Britannica* salesman arrived in answer to a mailed-in query (all babies eventually need an encyclopedia, don't they?) that I acted on the desire to learn about making yucca soap. We bought the *Britannica,* a purchase accompanied by the publisher's offer to respond to a certain number of research questions during our first year of ownership. The only question I ever asked was the yucca question. And in due time a fat envelope arrived, bearing two three-cent stamps and a postmark for September 8, 1956 (the baby was two weeks past her first birthday). In that day before computers and photocopies, the information requested, along with a twelve-book bibliography, was faultlessly hand-typed on five single-spaced pages. But, though I now had the facts on what to do, I stayed with Johnson's, and the yuccas in the now weedless garden remained intact. With reason: the *Britannica*'s research service had convinced me that only the roots of yuccas from the far West—from Arizona, New Mexico, and California—could be used in soap making. Nor did I experiment on eastern plants, or on any yuccas at all, until I was well advanced into grandmotherhood.

But meanwhile, I certainly thought about yucca soap, especially after returning to the Shenandoah valley town of my growing up. There, yuccas are planted ornamentally in gardens; they plant themselves in old fields and at the edges of woods; they even shoot up through cracks in the sidewalk. Everywhere, yuccas! And every June those beautiful white bells tremble atop tall flower stalks. I looked but did not touch.

The soap-making itch grew stronger, though not yet insistent enough to act on, when I came a dozen years ago to coastal North Carolina. Here yuccas volunteer with great exuberance; they rise out of beach sand, out of woods mold and pine straw, out of the very grass in our yards. Here, for the first time, I began to notice detail, to learn one species from another. The one that makes itself at home in every conceivable place is *Yucca filamentosa*—yucca "full of threads"—which forms a ground-hugging rosette anywhere from a few inches to two feet in diameter. Its pliant green leaves look frayed, with many fibrous grey-white strands curling outward along the edges. For these ravelings and the bluntly pointed leaf tips, this yucca is sometimes called Adam's thread-and-needle, or simply Adam's needle, a nickname it shares with at least one other member of the genus. Along the southeastern coast, it's more commonly called beargrass, perhaps because of its slightly shaggy appearance. The other locally abundant species is *Y. aloifolia*—"aloe-leaved" yucca—which is treelike, new leaves springing upward from a small, stout trunk. This is one of several yuccas with needle-tipped leaves that are known as Spanish dagger or Spanish bayonet. Watch out for mugging; beware of ambush. Once, inadvertently backing into a leaf, I felt nothing and was completely unaware that my lower calf had been wounded until friends alerted me twenty minutes later to a pant leg soaked with quiet blood. A third southeastern species, found in the salt air of many coastal venues from the Outer Banks to the Florida keys, is *Y. gloriosa,* a name that needs no translation at all. Going by the common name of mound lily, it can be readily distinguished from the other two by the thin red-brown marginal line on every leaf.

The mound lily is indeed an impressive yucca, a plant of "regal pomp and splendour." So wrote the American botanist William Bartram (1739–1823), who made an extensive study of southeastern flora (and anything else, natural and human, that captured his fancy). His account of these adventures, *Travels through North & South Carolina, Georgia, East & West Florida,* was sold by subscription and published in 1791; one subscriber was Thomas Jefferson, who paid sixteen shillings. It pulses to this day with Bartram's immense enthusiasm and excitement. And nowhere are these qualities more apparent than in his account of the glorious mound lily. Arriving at a Georgia sea island, he and his party pitched their tents "under the shelter of a forest of Live Oaks, Palms, and Sweet Bays," where they were much bothered by mosquitoes, roaring "crocodiles" (American alligators), and the seabirds coming and going by raucous thousands. And there on the shore, between forest's edge and the saltmarshes, was

> a barricade of Palmetto royal (Yucca gloriosa) or Adam's needle, which grows so thick together, that a rat or bird can scarcely pass

> through them; and the stiff leaves of this sword plant, standing nearly horizontally, are as impenetrable to man, or any other animal as if they were a regiment of grenadiers with their bayonets pointed at you. The Palmetto royal is, however, a very singular and beautiful production. [It] rises with a straight, erect stem, about ten or twelve feet high, crowned with a beautiful chaplet of sword or dagger-like leaves, of a perfect green colour, each terminated with a stiff, sharp spur. . . . This thorny crown is crested with a pyramid of silver white flowers, each resembling a tulip or lily. These flowers are succeeded by a large fruit, nearly of the form and size of a slender cucumber, which, when ripe, is of a deep purple colour. . . .

But for all that Bartram admired the rest of the plant, including its thickety habit of growth, he did not much appreciate the purple fruit. Though it is juicy, aromatic, and often eaten, he declared that it tastes bitter and "if eaten to excess, proves violently purgative."

Notice of the glorious yucca—and, indeed, the plant itself—had reached Europe nearly two centuries before William Bartram went on his wide-eyed trek through the Southeast. John Gerard gave the Old World its first description of *Y. gloriosa,* which he referred to as *Yucca Indica*—"of the Indies"—in his 1597 *Herbal,* a grand omniumgatherum in which accurate botanical information mingles intimately with Elizabethan fancies on the medicinal "Vertues" of plants and their "Temperatures," hot, cold, dry, or moist, according to the medical theory in fashion at the time. He constructed his description of the glorious yucca on the basis of personal observation. A specimen grew and flowered in his own garden, where "[i]t keepeth greene both Winter and Sommer . . . without any coverture at all, notwithstanding the injurie of our cold clymat." He also wrote of the plant's "very great root, thick and tuberous, and verie knobby, full of juice somewhat sweet in taste, but of a pernicious qualitie. . . ."

Imagine having your mouth washed out with soap. Imagine swallowing it. Though Gerard's account nowhere refers to soap, I think it likely that saponin, the suds-making constituent of the root, was responsible for this particular perniciousness. Juice from other parts of the yucca can be transformed, however, into potable beverages that only on occasion produce ill effects—but of this, more later.

The name "yucca" was placed on the plant by the Spanish adventurers who roamed the New World throughout the sixteenth century. And the Spaniards found the plant, in one species or another, almost everywhere, from Florida and the West Indies to the deserts of the Southwest, and on down into Mexico and the rest of Central America. How the Spaniards came to use the word "yucca" for this plant is a matter for speculation. In the September 1941 issue of a trade journal called the *Textile Colorist,* an article on yucca-derived saponin averred that the name was originally a Spanish word for "bayonet" and was applied to the plant because of its needle-tipped leaves (for this obscure tidbit, I bow to the *Britannica's* long-ago researches on my behalf). But I think that's putting things backward; the leaves suggested a pointed weapon that could have been called by any appropriate knife word from stiletto to Arkansas toothpick (had the Spaniards known that phrase). More telling, the word "yucca" is not native to Spanish. The 1948 edition of *Taylor's Encyclopedia of Gardening* may come closer to the truth: "*Yucca* is the Latinized version of a Spanish vernacular for some other desert plant." But it's Charles Sprague Sargent (1841–1927), a tree man of the highest order, who probably comes closest to the truth. In his splendid *Manual of the Trees of North America,* the second edition of which appeared in 1922, he said that the generic name *Yucca* "is from the Carib name of the root of the Cassava."

Yes! Though the Caribs may not be the source, the Spaniards must have adopted and adapted a term originally used by some indigenous people for a native plant. And it's entirely possible that the plant was indeed the tropical cassava. How so? In the first place, because of the close physical resemblance of the cassava to the yucca; both grow in rosettes, and both have stiff, narrow, needle-tipped evergreen leaves. Also, John Gerard complained a full three centuries ago that someone had tried to convince him that yuccas do not bloom, though his own garden had produced luxurious evidence to the contrary; the mistake, Gerard thought, was due to a mix-up between the yucca and the cassava—but the root of the former had the already noted pernicious effects, while the root of the latter provided an edible substance that Gerard called "Indian bread." The clincher is that, to this day, the starchy cassava root provides food, including tapioca, that is widely eaten by Indians living in the rain-forest villages of Ecuador and Amazonia; it's also used to produce a low-alcohol beer called *chicha.* (The brew is something for which I'm not at all sure that I could acquire a taste: the root is thoroughly chewed, spit into a container, and allowed to ferment.) Their word for the cassava plant is *yuca.* No wonder the Spaniards plucked the cassava's indigenous name and slid it over onto a plant of remarkably similar appearance. And there, on the yuccas, it stays.

And, oh, the yuccas that the Spaniards saw! Small yuccas and large, yuccas with edible fruit and yuccas without, yuccas with trunks, yuccas like great green rosettes, and a tree-tall yucca with twisty branches. The genus *Yucca* numbers forty species, all found only in the New World, and fifteen of these are native to North America. In addition to the thready, aloe-leaved, and glorious yuccas, several other species grow in the Southeast—Small's yucca, *Y. smalliana,* up in Virginia's Blue Ridge mountains; and the pale-leaved "grey" yucca, *Y. glauca,* a native of the Southwest that has been widely naturalized in the eastern states. It is in the Southwest—the sands of the Mojave Desert, the dry uplands of the Sierra Madre, the slopes of California's Tehachapi Mountains, elsewhere in arid places throughout the region—that yuccas most especially flourish. Some of these are the datil or "berried" yucca, *Y. baccata;* the "tall" yucca, *Y. elata;* and the Mojave yucca, *Y. schidigera,* the species name of which means "narrow leaved." The yucca with the name meaning "short leaved," *Y. brevifolia,* is the strangest of them all. Its branches reach out higgledy-piggledy like knotty, wildly contorted arms, with leaves sprouting from their tips like stubby green fingers. This is the

Joshua tree, so named by the Mormons, who thought that it seemed to beckon them onward through the wilderness as Moses' successor Joshua had beckoned the Israelites to follow him toward the Promised Land. Though long considered part of the Liliaceae, the Lily family, these yuccas and all the others are now classified by most botanists with the Agave family, the Agavaceae. The plants for which this family is named include the slow-blooming century plant, *Agave americana,* and the fibrous plant from which sisal is made, *A. sisalina.*

The yuccas may be the most peculiar members of the Agave family—at least, in matters of reproduction. They need their moths. Although they are able to multiply (in good plant fashion) by sending up shoots from runners, they cannot produce seeds without help from a moth, and not just any moth but one especially adapted to each of the forty species in the genus *Yucca.* And these moths—uniformly pallid, unobtrusively small—depend equally upon beargrass, mound lily, Spanish bayonet, Joshua tree, or one of the others to keep their own kind going. In a completely benign relationship known as mutualism, insect and plant are each vital to the life of the other. The genus name of the moths, *Tegeticula,* tells part of the story: it means "little concealer" or "little burier." That's an apt way to describe the female moth's instinctive habit of making multiple punctures in the pistil of a yucca blossom, putting one tiny egg into each tiny hole, and filling the remaining space with balls of pollen, which she rolls and carries with special tentacles near her mouth. The results: a pollinated flower that produces seeds—seeds for the moth's finicky, one-food-only caterpillars to eat and seeds for the next generation of yuccas.

Some of the yuccas also supply food to humankind—food of both solid and liquid sorts. The former, immemorially part of the Native American diet, includes the large, yellow, banana-like fruit of the datil *(Y. baccata)* and the smaller red-brown fruit of the Joshua tree, which has sometimes been ground into meal as well as eaten raw. But edibles like these are small-time compared to a notable beverage, tequila. Hoist it up and drink it down, this yucca-based product must be equally immemorial in its use, which is hardly confined to indigenous peoples but has made friends and founded Margaritavilles around the world.

Tequila is made from various members of the Agave family, notably the yuccas. To qualify in the first place as a drink legally entitled to the name tequila, it must contain at least fifty-one percent agave juice. And the new premium tequilas, popular of late, hit the top of the scale with an agave-juice component of one hundred percent. The plants that yield the juice are not harvested in the wild but grown in plantations, and the juice, which will be fermented and distilled, is not the stuff about which John Gerard complained. Tequila's basic ingredient is made from flowers rather than saponaceous roots. Granted, flowers transformed into tequila may have qualities as pernicious as those of the roots—not, however, for all who partake but only for those who overindulge (peace, all ye who believe that alcohol is intrinsically poisonous).

And there's still no end to the traditional uses of the yucca: the plants have not only provided humankind with food and intoxicating drink, but the leaves of many species yield fiber strong enough for hats, mats, baskets, and ropes. Archaeologists have discovered woven yucca-leaf baskets dating back to about 300 B.C. in a New Mexico cave. And other parts of the plant are just as useful. The trunks of southwestern yuccas have been used as fence posts and even as walls for houses. A dye has been extracted from the roots of the Joshua tree. Nor can I possibly overlook the real suds, the magically cleansing lather—amole, the Spaniards called it, from a Nahautl (Aztecan) word meaning "soap"—that is the roots' primary gift. (And I'll tell you in a paragraph or two of the resolution to my quest.)

Ever since the days of John Gerard, the Western world has fitted the yucca to its own peculiar purposes—or not peculiar, as the case may be, except that our modern purposes do not speak to the basic human needs for food, clothing, and shelter. We use the plants not only for making a form of white lightning but also for decoration. Because they are pleasing to look at—oh, those great panicles of big, creamy, thickly clustered bells!—yuccas have found a place in the garden. Garden catalogues offer varieties developed in the last few decades, such as a variegated *Y. filamentosa* and another of the same species with leaves that are golden, not green. And New Mexico has selected yucca in full blossom as its state flower.

Nor do contemporary uses of the yuccas end with drinking them or planting them for show and ornament. In the current efforts to preserve or restore wetlands on the mid-Atlantic coast—or even build new ones—while at the same time allowing for some development, yuccas may come to play a valuable part. Wetlands, of course, have enormous ecological significance as necessary habitats for many kinds of wildlife and, thus, as sources for human recreation from boating and bird-watching to hunting. They serve, too, as filters for pollutants, buffers against erosion, and great sponges that act as natural flood-control devices. One of the secrets to re-creating a wetland is to provide the appropriate flora—common marine plants like eelgrass and widgeongrass, everyday marsh and beach plants like spartina and yucca. But because of the potential for introducing diseases and insect pests, it's not advisable to dig plants elsewhere and import them to the site. Aside from which, raiding one wetland to build up another is a self-defeating exercise. Science to the rescue: a never-ending supply of clean, healthy specimens may be cloned in a laboratory for setting out later in the wild. This enterprise is still in the planning stages, but at this moment, in a University of North Carolina lab, little yuccas grow and provide material for propagating more yuccas and still more yuccas and . . .

All the yuccas I need grow right in my yard. And my try at making soap arrived in the spring of my sixty-third year, mere weeks after I'd read in an herbal that *Y. filamentosa,* the yucca most common in our area, has a saponin content of two percent. That's a far cry from the western yuccas, which may contain from six percent saponin to a walloping fourteen percent. But, I was assured, saponin lay close at hand—if only I took the time to get it. The long-postponed moment came along this way:

"That yucky has got to go," said my husband, the Chief, pointing at the large, frayed-looking Adam's thread-and-needle located two feet from the back steps. Though it had bloomed vigorously, gorgeously just last year, he's never liked it. A man who wants plants to grow where he puts them rather than in any old spot of their own choosing, he's mowed it down at least a dozen times. He knows, however, that it will shoot right up again. He also knows that I like it.

"Go? For heaven's sake, *why?*"

When asked that kind of question, he's apt to answer simply, "Because." But this time he supplied an augmented version of the word. "Because I am going to build you a back deck—a big one, eight by sixteen—and it will cover that damned thing right up."

A back deck—excellent idea! The back of the trailer receives shade from midafternoon on. The front deck is a friendly place till then, but when the sun starts westering, its light bounces off the river, and not all the pines and sweet gums in the front yard, not my dark glasses, not a baseball cap with a big brim offer respite enough from glitter and blinding glare.

"Build!" I said.

Imagination insisted that the yucca would be hard to dig up, that it and all of its kind were anchored in the soil by a system of interlocked and clinging runners, which would staunchly resist my attempts to uproot them. And hadn't the information supplied decades ago by the *Britannica* said that it took not just pick and shovel but a crowbar to get the deep-seated roots out of the ground?

That may be true for treelike western yuccas but not for the low-growing rosette by our back steps. I used a spade to make a circular cut eight inches deep around it. The only obstacle encountered was trifling—the edge of a pine root. A light heave on the spade, and the yucca was lifted free. I trimmed off the leaves then, and gave the root a good scrub to remove clinging earth.

What a lumpy, bumpy root it was—a Siamese potato of a root, one bulbous section the size of my fist joined to another of similar proportions. It smelled as fresh as a newly dug potato, too. And its flesh was just as white and crisp. I retrieved the *Britannica*'s information sheets from the bulging file box in which they'd long languished with other potentially useful bits of paper. They suggested several methods by which yucca roots, be they newly harvested or dry, might be prepared for yielding soap: slicing, grating, mashing with a hammer or a stone. I fetched my kitchen grater, the old-fashioned, hand-held kind that has four different cutting surfaces. With grater, plastic pan, and fine, knobby root, I went outside to the cool May-morning shade of the front deck and set to work.

The root shredded easily. And as it was rubbed over the grater's next-to-largest holes, it left faint but unmistakable streaks of whitish lather on the metal. With a finger, I wiped off a trace and tasted it. Ugh, definitely saponaceous. The moment of testing came when the plastic pan held a cup's worth of white shreds: time to mix them with water. The long-saved information mentions the use of both hot and cold water. I opted for a quart of lukewarm. And stirred the mix with a vigorous hand. And produced suds. They foamed and rose. They billowed above the pan's rim. There was no slippery, soapy feel to the liquid, but its scent was delicate and enticing, fresh as grass, clean as sunlight. I happily sloshed the suds up my arms and over my face.

Since then, the back deck has been constructed. Late afternoons, I sit reading in a rocking chair placed over the spot where the yucca used to grow. It's not entirely gone, however. Only a quarter of the root was used. The rest has been placed in a string bag lest it mold, and it is stored with other plant materials, like sun-dried basil and sassafras roots. I'll shred more the next time I long to be bathed in pure satisfaction.

ERASMUS DARWIN AND THE UPAS TREE

An Interlude in Other Places, Other Times

■

The most famous curiosities amongst Asian trees are the Upas or Ipoh (Antiaris toxicaria) *and the strangling and banyan figs. The former is, with strychnine, the main constituent of arrow poisons and knowledge of it was long withheld from Europeans; to forestall inquiry the myth was propagated that it killed all that came near it, a fantasy which gained extensive currency in Europe. It has a wide range throughout tropical Asia and in many forests is rare, but it is never found without tapping scars.*

The Oxford Encyclopedia of Trees of the World

Clearly, the upas has nothing to do with the arborescent flora of North America—nothing direct, that is. But this curious tree has a curious history: it underwent a sort of mutation that resulted in a remarkable new genus and species. The mutation is indeed so different from *A. toxicaria,* its still surviving progenitor, that it may even be placed in a different, and most ancient, family, the Fabulaceae, which has been found in every land since people started paying close attention to natural phenomena. It was at one time possible to count the mutant offshoot as a living member of the Fable family; indications are that it was an evolutionary failure. It is now extinct and has probably been so for nearly two hundred years—but not before gaining some international notice in the later part of the eighteenth century. Today, however, only a pair of botanical references from that time provide the main evidence of the tree's existence. Although these descriptions, one in Dutch, the other in English, are sketchy at best—and the latter written in an altogether original fashion—taken together they make an excellent case for what might be called "earned" extinction. The mutant upas was not something a reasonable person would plant for shade or ornament nor something even the most unreasonable would want to stumble on. It was an unfriendly tree. More than that, it was often fatal.

We may thank the Dutch report for primary information about habitat, appearance, growing conditions, and human uses of the tree's main product. Written in the late 1770s by one N. P. Foersch, a surgeon employed by the Dutch East-India Company and stationed in Batavia (modern Jakarta) on the island of Java, it was translated soon thereafter into English and published by *London Magazine* in either 1783 or the year following. The description first given—or, more precisely, composed—in English appeared in 1789. It is the work of Erasmus Darwin, physician and botanist extraordinaire (other adjectives also apply and shall, in due course, be mentioned). Though Dr. Darwin's report is secondary and quotes in its entirety the earlier account by Mijnheer Foersch, it may fairly claim credit for making knowledge of the upas tree available to our modern attention. But let me take these two documents one at a time, in the order in which they were presented to the world.

Foersch's account was based on two sources: his personal, if somewhat distant, acquaintance with the tree and the words of native informants. For reasons that will soon be evident, it was necessary to rely on the latter because Foersch himself dared not go close enough to the tree to see it plainly, much less take measurements of its height, crown, and girth. But from his findings (and from the translation in *London Magazine*), a brief botanical description may be constructed. The name by which the tree was commonly known is followed by a version then in local use.

Upas
"Bohon-Upas"

Description: A tree of middling size that secretes a camphorlike resin, emits a lethal vapor, and forms a small colony by sending up root sprouts.

Habitat: Sandy soil of a brown color, strewn with stones and skeletons.

Range: One location on the island of Java, 27 leagues from Batavia, beside a rivulet in a mountain-encircled valley barren of other life, both plant and animal, for 15 to 18 miles in every direction.

The resin, or gum, is the upas tree's chief product. Because of the tree's fumes and the poisonous nature of the gum, collecting this substance is a task reserved for condemned criminals, who are given an unconditional pardon if they survive. Survival depends on walking when the wind blows toward the tree; the chance for a favorable breeze, however, is one in ten. The gum successfully collected is applied to the tips of the sharp weapons used to execute malefactors, such as the thirteen concubines convicted of infidelity to the Emperor's bed. There is only one other member of this genus, the Cajoe-Upas found on the coast of Macassar.

And these, according to Foersch, are the "simple unadorned facts, of which I have been an eyewitness."

If only the winds had favored a closer approach, if only he had been able to make hands-on examination of living buds, flowers, fruit, and leaves, he might have been able to adorn his tantalizing, far-from-simple facts with details. Though one of the criminals sent to harvest gum managed to return with two leaves, they were dried and evidently failed to thrill Foersch into an act of description. He did not leave posterity any account of their appearance, and we shall never know if they were toothed or lobed linear, spatulate, or oblanceolate, glabrous or tomentose (botanists are as fond as lawyers of exactly denotative polysyllables, but Foersch was neither—he was a surgeon). Later, however, he did make a full disclosure of the results obtained after he'd cadged enough gum to make predictably deadly experiments on several puppies, a cat, and a chicken. But Foersch's facts are generally more notable for what they omit than for what they include.

Nonetheless, I think it possible right here and now to fill one large hole in the surgeon's description (though it may be a hole of which he was completely, forgivably unaware). Botany's devoted amateurs will note that the brief description above makes no mention of the tree's binomial. In all likelihood, this faraway mutant, certainly already endangered but not then extinct, was never brought to the attention of the late eighteenth century's classifiers who had adopted the double-barreled Linnaean system, then brandnew. I take it on myself, therefore, to publish a name for this hitherto overlooked species: *Upas darwinii.* The genus name has been assigned, of course, because it was the term by which the tree was known to the inhabitants of Java (just as *Quercus, Salix,* and *Acer*—common names used by the Romans—now designate, respectively, genera for oaks, willows, and maples). For the specific name, I might have acknowledged the surgeon who first made the mutant upas known to the Western world, but I choose instead to honor the man who has managed, if minimally, to keep its memory alive.

Erasmus Darwin. It may be instructive to look at the man before quoting his report on the upas. His surname is not unfamiliar to students of science. Indeed, this Darwin, a physician by vocation, was responsible in more ways than one for the renowned naturalist who put together *The Origin of Species.* To begin with, Erasmus was grandfather to Charles. They never knew each other in person; Erasmus, born in 1731, died in 1802, seven years before Charles was born. But in the same way that the two shared a blood tie, some of the grandson's scientific thinking was directly descended from seminal hypotheses advanced by the grandfather. In *Zoonomia; or, The Laws of Organic Life,* issued in 1794 and 1796 in two volumes, Erasmus

Darwin presented his views of zoology—and his thoughts, as well, on the physical sciences, medicine, philosophy, religion, and almost everything else that had entered the sphere of his questing intellect and omnivorous curiosity. This treatise includes little material on botany, but Darwin more than made up for the lack by devoting at least four other volumes to that subject.

So it was that, along with his grandfather's genes, Charles Darwin inherited many ideas deserving much investigation. Erasmus Darwin had posited, for example, that the male animals and birds best able to fight off rivals were the individuals that "propagate the species, which should thence become improved" (*Zoonomia I*, 503). And he offered observations to suggest that individual survival depends on factors like being able to evade enemies and making the best use of an existing food supply, and that these abilities may also be transmitted as improvements to the species. But many of the ideas put forth in *Zoonomia* were considered radical and dangerous—indeed, altogether ungodly—for Dr. Darwin had dared to support the blasphemous notion that life is subject to change rather than having been immutably fixed from the moment of Creation. This proposition so offended conservative scholars and prelates that the book not only drew furiously hostile criticism but also earned a place on the Catholic index of forbidden books.

His work on botany might have found a similar niche, except that Dr. Darwin found a novel strategy to disarm his critics' wrath: clothing science in the garments of art—of poetry, to be exact. His first grand foray into the subject was *The Botanic Garden*, for which William Blake engraved some of the plates. With its second part published before the first in 1789 (the year of the outbreak of the French Revolution) and the whole appearing in 1791, it consists of two volumes of heroic couplets, wrought (some might say overwrought) after the fashion of Alexander Pope. The rollicking, rhyming iambic pentameter of the main text is amply supplemented, however, by "philosophical notes," as Darwin called them—prose footnotes in minuscule print. His third volume on botany, *Phytologia*, published in

1800, lapsed entirely into plain, straightforward prose, but *The Temple of Nature,* which appeared posthumously, returned to rhyme.

Why *poetry?* Darwin himself presented a reason in Volume II of *The Botanic Garden,* the volume entitled *The Loves of the Plants,* which contains the lines on the upas tree and the other Darwinian passages quoted here. In this book, a prose "Interlude" separates each "Canto," or main section, from the next. These Interludes are constructed as dialogues between a Bookseller and a Poet. They begin with a skeptical, almost belligerent comment from the Bookseller: "Your verses, Mr. Botanist, consist of *pure description,* I hope there is *sense* in the notes." Drawing on examples from all the arts, citing actors, painters, playwrights, and poets from Homer to Shakespeare, the Poet proceeds to show the sense in using verse. Through poetry, he says, "the irritations of common external objects" are banished and a "reverie" is induced, in which the subjects of the verse become objects that "appear to exist before us." The Bookseller makes instant retort:

> *Bookseller:* It must require great art in the . . . Poet to produce this kind of deception.
> *Poet:* The matter must be interesting in its sublimity, beauty, or novelty; this is the scientific part; and the art consists in bringing these distinctly before the eye, so as to produce the ideal presence of the object. . . .
> *B.:* Then it is not of any consequence whether the representations correspond with nature?
> *P.:* Not if they so much interest the reader or spectator as to induce the reverie. . . . The further the artist recedes from nature, the greater novelty he is likely to produce; if he rises above nature, he produces the sublime. . . . (56)

Aha! The elder Darwin, by his own admission, intended to produce not only a supernatural sublimity but also deception. It was not, however, the kind of deception about which the Bookseller questions him. The Bookseller, a fictive device, served handily as Darwin's foil to mislead the critics. Not his literary critics; I doubt that Darwin cared a whit about their comments, including Samuel Taylor Coleridge's rude remark, "I absolutely nauseate Darwin's Poem." The trickery practiced here was meant to disarm the traditionalists who would attack his radical, apparently irreligious ideas about nature and human nature. In other words, Erasmus Darwin could most slyly shield himself against their slings and arrows, could justify anything he chose to say simply by claiming poetic license.

The Loves of the Plants, the volume with the lines about the upas tree (patience; I'm getting there), is radical indeed. It aims to show the myriad reproductive arrangements to be found, and observed, throughout the vegetable kingdom. More important, it is intended to support Darwin's budding theories on reproductive rivalry as a component of species improvement. Completely genuine in its scientific purpose, *The Loves* is based on the first system of plant classification devised (but later rejected) by Linnaeus, whom Darwin greatly admired as "the Swedish sage." Sometimes called the sexual system, it organized plants into classes according to such criteria as the perfect or imperfect nature of their flowers and the number of pollen-bearing stamens in a single flower. The fact that plants have sex lives was still something of a shocker; the idea had gained scientific credence only at the beginning of the century. Darwin's poems and philosophical notes explain this system with specific, intimately detailed examples of the plant world's amazing versatility in matters having to do with acts of reproduction.

The Loves of the Plants may well have titillated scores of readers, may well have lured and seduced them into not only reading but buying: Its couplets are the botanical equivalent of *The Joy of Sex.* And how sublime, that a reader might revel in such high-flown erotica on the ground that it was sincerely educational! Our botanist was not only extraordinary but also—here come the adjectives promised earlier—voyeuristic, prurient, and grinning with sheer impudence. But then, according to an anonymous obituary, the good doctor was known to the end of his days as a man who sacrificed often and most affectionately to Venus.

Darwin rationalized the fashion in which he presented his views by saying that if Ovid-could use poetry back in the days of Augustus Caesar to "transmute Men, Women, and even Gods and Goddesses, into Trees and Flowers; I have undertaken by similar art to restore some of them to their original animality, having remained prisoners so long in their respective vegetative mansions" (x).

The animality of plants! Working the classic metamorphosis in reverse, Darwin turned plants into animals and, more exactly, into animals of the human kind. Examples of his method are certainly called for. The first describes the foreplay engaged in by *Tropaeolum majus*, the common garden nasturtium, which had been reported by Linnaeus's daughter to emit phosphorescent flashes just before sunrise. The number *eight* refers to the number of stamens standing ready to shed their pollen on the flower's single pistil.

> The chaste TROPAEO leaves her secret bed;
> A faint-like glory trembles round her head;
> *Eight* watchful swains along the lawns of night
> With amorous steps pursue the virgin light;
> O'er her fair form the electric lustre plays,
> And cold she moves amid the lambent blaze. (148–9)

The second example details the sexual behavior of a sensitive plant that Darwin knew as the chunda, an herb more formally known as *Hedysarum gyrans*, "sweet fragrance that whirls," a member of the Pea family. Again, the italicized number refers to the plant's aroused stamens, which here bend themselves eagerly over the pistil.

> Fair CHUNDA smiles amid the burning waste,
> Her brow unturban'd, and her zone unbrac'd;
> *Ten* brother-youths with light umbrella'd shade,
> Or fan with busy hands the panting maid;

Loose wave her locks, disclosing, as they break,
The rising bosom and averted cheek;
Clasp'd round her ivory neck with studs of gold
Flows her thin veil in many a gauzy fold;
O'er her light limbs the dim transparence plays,
And the fair form, it seems to hide, betrays. (172–3)

Anthropomorphic, yes, but with reason: What better way to transform stamens and pistils, pollen and ovaries from necessary but uninteresting parts into subjects of fascinating, page-turning scandal? Here Chunda is clearly, willingly poised for a gang bang.

Afterward, however, the plant may sleep. Add credulous and tender to Dr. Darwin's adjectives. The philosophical note accompanying these lines describes the continual rising, falling, and whirling of chunda's leaves, even when the air is still and warm. This description then segues into a meditation. Such movement, such "vegetable spontaneity," also occurs in other plants. But when movement ceases, it does so because the plant has suspended voluntary action and, like any animal, has gone to sleep. But, are these motions truly voluntary on the part of any plant? Darwin answers yes, "for without the faculty of volition, sleep would not have been necessary to them."

It's time, past time, to head back to Java and the unique upas tree. Here, without further delay, is Erasmus Darwin's report. Note that this exotic mutation, though fully endowed with animality, is not the sort to engage in panting dalliance. It evoked admiring horror rather than an examination of its sexual proclivities.

The report begins with a description of the scene: a rocky, windswept plain ruled by eternal summer. Though rain falls abundantly here and streams flow, no nutmeg and plantain, no grass, flowers, mosses, and lichens find a foothold, no fish swim in the streams; no birds wing through the air, neither moles nor worms mine the soil. Though the ground is littered with bony evidence that various creatures have come this way, no living thing may exist here—except for one:

Fierce in dread silence on the blasted heath
Fell UPAS sits, the HYDRA-TREE of death.
Lo! from one root, the envenom'd foil below,
A thousand vegetative serpents grow;
In shiny rays the scaly monster spreads
O'er ten square leagues his far-diverging heads;
Or in one trunk entwists his tangled form,
Looks o'er the clouds, and hisses in the storm
Steep'd in fell poison, as his sharp teeth part,
A thousand tongues in quick vibration dart;
Snatch the proud Eagle towering o'er the heath,
Or pounce the Lion, as he stalks beneath;
Or strew, as marshalled hosts contend in vain,
With human skeletons the whiten'd plain.
—Chain'd at his root two scion-demons dwell,
Breathe the faint hiss, or try the shriller yell;
Rise, fluttering in the air on callow wings,
And aim at insect-prey their little stings. (115–6)

Upas darwiniia—a tree that hisses! A tree with teeth! A many-crowned tree with sturdy offshoots that hone their deadly skills on insects before they attempt eagles and lions. A tree entirely worthy of Erasmus Darwin's questing curiosity and his name.

The mutant upas is also a tree that perfectly represents the toxic members of its family, the Fabulaceae, at which it may be well to take a brief look. The defining characteristics of the family as a whole are animality and volition—that is, behavior closely resembling that of human beings, along with an ability to exercise choice. Within that general frame, however, the family's members fall into two distinct groups—the harmful and the benign, each with more genera than you can shake a stick at. Two such trees, one from each group, have been notably published in *The Lord of the Rings*. (In botanical parlance, "published" means "first given full description.") Travelers through the Third Age of Middle-Earth will recall them well.

The first, a fairly murderous specimen, is *Archaealsus salignosenex* (again I take the liberty of assigning binomials). The genus designation means "Old Forest," the location in which the type specimen was found; "old man willow" is English for the species name. The nature of this tree and those that share its genus is best revealed by the man who discovered and described them, J. R. R. Tolkien. Examining "the hearts of [these] trees and their thoughts, which were often dark and strange, and filled with a hatred of things that go free upon the earth, gnawing, biting, breaking, hacking, burning: destroyers and usurpers," he concluded that "[t]he countless years had filled them with pride and rooted wisdom, and with malice." As a result, these ancient trees, especially *A. salignosenex,* exercised their inbred volition and animality by attempting, and often succeeding at, the entrapment and destruction of all free-moving creatures that enter their domain. There is no current evidence, however, that any members of this genus still exist in the wild.

The benign arborescent species described by Tolkien is almost certainly extinct. It was already, by its own acknowledgment, on the way out at the time that he discovered it. I've chosen to call it *Entus tolkiensis,* which combines a latinization of the name it gave itself, The Ent, with a species designation honoring the amateur botanist who announced it to the world. Note that one of The Ent's distinguishing features was the power of speech; not only did it pronounce its name and prophesy its disappearance, but it denied that it was a tree. Nonetheless, Tolkien has made it clear that *E. tolkiensis* resembled a wind-sculptured bristlecone pine—"one old stump of a tree with only two bent branches left: it looked almost like the figure of some gnarled old man." He also made a distinction, worthy of Erasmus Darwin, between the sleeping Ent and The Ent wide awake: the former was all tree, at least to the casual observer; the latter, a tree that not only talked but walked and provided transportation services for the creatures cradled in its branches.

Like these two species, the kin of the upas have exercised their goodwill or malevolence since humankind climbed out of its own tree—and perhaps even before then. Think of the Grimm forests, those ideally crepuscular habitats for wolves and witches, where branches have ever reached out to snag and hold unwary travelers. An equally sinister forest, growing midway between the Emerald City and the Land of the Quadlings, was reported in 1900 by an American, L. Frank Baum (whose surname means "tree" in German); the trees on its edge employ their supple branches to seize intruders and fling them back. Think, too, of vines, for the family Fabulaceae contains nonarborescent members, such as Jack's beanstalk. (Do not, however, think of the laurel that was once a nymph named Daphne nor of Yggdrasil, the sacred ash tree that supports the universe; these are members of another family altogether, the Mythologaceae.)

The trees listed above are primarily Old World species. But North America is not completely without a small share of native Fabulaceae. One of the family's minor representatives is confined to the habitat provided in the neighborhood of the Big Rock Candy Mountain: the cigarette tree, which forever invites buzzing bees to make honey from its blossoms and actively entices all who pass by to help themselves to the tidily rolled fruit (note, however, that the fruit now bears a warning from the surgeon general). This species is relatively harmless, however, in comparison to the New World's night tree (I take the liberty of designating it *Arbor noctis*), which exhibits as much malevolent animality as the upas. It is most graphically described in Conrad Richter's book *The Trees,* first published in 1940. It gives a realistic account of the trials experienced by Americans in the late eighteenth century as they pushed westward through the endless and shadowy old-growth forests of southern Ohio and adjacent parts of Kentucky. The night tree becomes active only after sundown. Then it can be heard hissing like snakes and knocking on the doors, walls, and roofs of isolated cabins. Its aim is not to keep people out of its domain but to trap them and forever hide all traces that they ever existed. On the narrow escape of a small party rescuing a girl named Genny, Richter writes:

> All the way home the woods lay dark and dripping. The heavy butts of the trees nearest the path moved furtively behind them as they tramped, but the furthest ones stood watching them go. Oh, those wild trees stood stock still like they hated to see Genny pulled out of their clutches. They thought they had her fast. . . .

Nowadays, though *A. noctis* still thrives, still knocks and hisses in less traveled regions, its range has been much reduced.

I cannot close this brief survey of the upas and some of its relatives without an ultimate nod to Erasmus Darwin. He is, after all, the man who inspired my investigations. And because he has put the thought so neatly, I shall relay to the reader the final remarks made by his Poet to the Bookseller. Placing both volumes of *The Botanic Garden* in the Bookseller's hands, the Poet says with a smile and a shrug:

> I now leave it to you to desire the Ladies and Gentlemen to walk in; but please to apprize them, that, like the spectators at an unskillful exhibition in some village-barn, I hope they will make Good-humour one of their party; and thus themselves supply the defects of the representation. (144)

The only thing left to be said is that imagination has always been earth's most fertile soil, the rich ground in which sense, as well as nonsense, must take root before it grows and thrives.

DAPHNE

Well-rooted, glossy, thriving, you
arrive by second-day delivery, UPS,
ready for planting (if climate's mild)
or simply setting out, still in familiar
potted earth, in window or on front porch
beside geraniums and three varieties of parsley.
And you are now as willing as you were then—
except for that one swift, staggering
act of seizure—willing to let yourself
be shaped in many guises, cone or star, square
balanced on one tip, perched bird, good
girl, obedient and eternal daughter.

The instant your father the river
bent you, broke you to his will and forever
stopped your feet from dancing or flight, what
did you feel? Stasis must have struck you
like a stunning blow, wood suddenly
subsuming flesh. But in that final
gasp of breath before fingers leafed out,
before sap ran in your veins and your heart
hardened into dead fiber, did anger flare,
or regret? Or here-and-gone gladness
that not even a god, aroused, erect, could
tear, could penetrate a tree?

Laurus nobilis, celebrated laurel, sweet bay,
sweet girl, fragrance and seasoning in one—
and in many, $3.50 to $8.50 each
in pot, price depending on the nursery,

and each of you in direct descent
from that arresting transformation, each
a nymph in your own right, safe
in rooted stillness and hallowed
by your own green life. The comfort here:
outlasting death. As for the proud,
careless gods, not one can be had these days,
not for love nor money.

THE DEVIL'S WALKING STICK

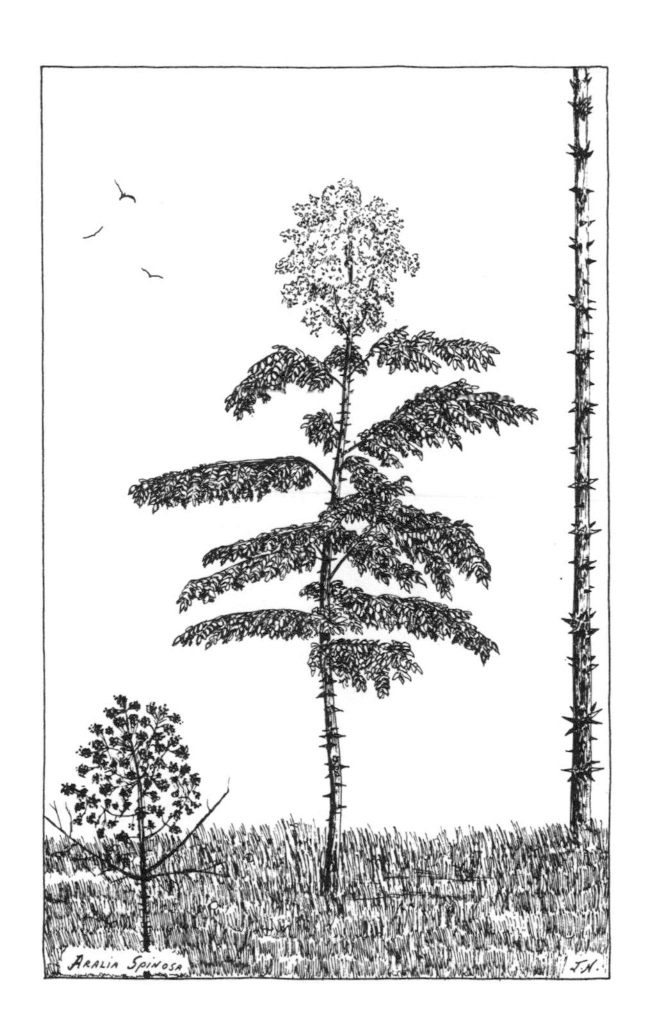

Portraits of the Devil depict many aspects of that master of lost souls. They show the horns on his dark head and the twitching, hairless tail with an arrow at its tip; they capture the skepticism in his raised eyebrows, the feral shining in his coal-black eyes, the eternally gaunt and greedy hunger of his smile. Sometimes they reveal the hooves that he usually conceals within his patent leather shoes (or, as the case may be, his snakeskin cowboy boots). But they don't show every feature, no indeed. Take, for example, his lean, long-fingered hands. Whether they clutch the emblematic pitchfork or grasp the contract some poor mortal has signed in blood, only their backs are visible. What of the palms? Has anyone ever made a drawing of the Devil's palms? Or even looked at them?

I've certainly never seen them (nor the Devil himself, so far as I know), but I have a fair idea of what those palms must be like. They're hard as iron, thick as elephant hide, unfeeling as horn. They have to be. Else, how could the Devil, his fearful powers notwithstanding, bear to get a firm grip on his walking stick? And that I *have* seen.

If His Satanic Majesty were to stroll down the Atlantic seaboard from New Jersey to central Florida, then saunter along the Gulf coast into east Texas, and complete his perambulations by heading northeast across Missouri and on over the Appalachians through Kentucky and the Virginias till he returned to his starting point, he'd find a live, new walking stick every time the old one wore out. Country byway or interstate highway, an abundant supply of walking-stick trees grows on shaded stream banks and at the edges of the hardwood forests that border many southeastern roadsides. It's not that the Devil needs a support, however, to assist his infernal progress—no flat feet, no wonky back or arthritic knees for him. I'd bet that he uses his stick to show off, throwing it and catching it as if it were a song-and-dance man's cane, twirling it through his fingers like a baton, and sometimes closing his hand tightly around it to inflict a few whacks on his latest victim. A painful experience: the stick bristles with spines.

Science, omitting mention of the Devil, focuses on these many spines. Sharp as needles, they thrust not only from the bark but also from the tender green leaf stems of a most peculiar tree that is found only in North America. Its formal name was selected back in the mid-eighteenth century by Linnaeus himself. That sensible man sorted through the descriptions made by naturalists in the field and chose *Aralia spinosa.* Today no one knows how and where the genus name *Aralia* originated nor what it means, but the Latin word *spinosa* says today what it's always said—"full of spines."

These facts of nomenclature did not, however, offer themselves to my attention until late in the sixth decade of my life, well after I'd come to live for eight months out of every twelve on the shore of North Carolina's wide and salty river Neuse. The tree grows abundantly here, and I could have encountered it all on my own. But for five whole years, I ignored the species. Not only ignored it, but failed to see it in the first place. I didn't see the many spine-embellished walking sticks that rise along the first three miles of the road to town or, shamefully, the few that grow at the edge of our yard, right beside the path I usually take when I head for the woods. It was my husband, the Chief, who finally performed the introduction on a late-winter morning.

His shout brought me out of the house at full speed: "Hon, there's something funny out here!" I've learned to make immediate response to that holler; issued not infrequently, it tells me that something interesting, unfamiliar, or downright peculiar has just appeared on the natural scene. He's hollered about a newly hatched luna moth clinging to the grey bark of a pine to flex and dry its luminous pale-green wings. He's hollered about a great cormorant, a first for the county records, catching flounder out there in the river. He's hollered about golden chanterelle mushrooms pushing through leaf mold into the light—and into my skillet. And he's hollered about a gargantuan orange caterpillar galloping across our drive, its rump decorated with the horn characteristic of all Sphinx moth larvae, its body the size of a fat cigar. His cry may alert me to something as dramatic as a young bald eagle perched in a riverside cypress only four hundred feet away—or to something as grotesque as the plant he pointed out that winter morning. "Hon, what the hell is *that?*"

It certainly didn't look like a tree. Slender as a rake handle, straight as a yardstick and not much taller, its entire length studded with row on irregular row of short, needle-sharp, no-nonsense spines, it looked like a botanical experiment gone wrong. But nearby stood two others of its crazy kind, all growing unobtrusively amid wax myrtles and yaupon hollies at the edge of a pine and sweet gum grove. They had no branches; instead, each sported a single large bud at its skyward tip. No more than a brief consultation of a field guide was needed to place both binomial and not one but several common names on these trees ubiquitous throughout the South. "Do we really want those durn weird things in our yard?" the Chief asked. I told him yes, that weirdness exerts its own fascination. He did not fetch his chain saw.

Waiting for spring, longing for light, warmth, the return of nesting birds, and the sprouting of buds (especially those atop the Devil's walking sticks), I took the edge off impatience by looking into *A. spinosa.* It's one of the Araliaceae, the Ginseng family, which comprises trees, shrubs, woody vines, and perennial herbs. And like a wonderment of other families, its members are variously native to North America and Asia, but not to Europe and Africa. While the family exhibits a proper orderliness in respect to the botanical characteristics shared by its members (five epigynous stamens, ovaries inferior, that kind of thing), it is otherwise splendidly higgledy-piggledy in habit and appearance. Some are spiny—"armed," as the botanists would put it—while others are not. Among the family's unarmed perennials are the several species of ginseng, English ivy, and wild sarsaparilla. The last, a plant of rich, open woodlands, belongs to *A. spinosa's* genus, and its formal name, *A. nudicaulis,* speaks of its undefended state—"bare-stemmed" Aralia.

A reputation for medical potency adheres to a host of the Araliaceae. The aromatic roots of wild sarsaparilla provide a bitter-but-good-for-what-ails-you principle once used as a flavoring and stimulant (though another plant, *Smilax,* was the source of the tonic popularly known as sarsaparilla). A brew said to relieve rheumatism and ease the pains of childbirth has been decocted from the roots of the tall woodland plant *A. racemosa,* commonly known as American spikenard. And to this day ginseng of both the Asiatic and American sorts is famed as a panacea, revered as an aphrodisiac. But what, if anything, might *A. spinosa* be good for—apart, that is, from serving as the Devil's walking stick and magic wand? Quelling certain sorts of pain, that's what, and inducing sweat and—but let John Lawson outline some of the rest. When he published his admiring account of Carolina's wildlife and plants in 1709, he included this description of the tree he calls the "Prickly-Ash." (It's one of two trees known by this name, but of that, more later.)

> Prickly-Ash grows up like a Pole; of which the *Indians* and *English* make Poles to set their Canoes along in Shoal-Water. It's very light, and full of Thorns or Prickles, bearing berries in large Clusters, of a purple Colour, not much unlike the Alder. The Root of this Tree is Cathartick and Emetick, used in Cachexies.

In other words, people who were cachexic—wasting away with chronic disease or suffering persistent malaise in body or mind—could dose themselves with tea made from the tree's roots, effect a complete purge of stomach and bowels, and feel like facing the world again.

In addition to prickly ash, the tree bears two other common names that it shares with another species from another family: toothache tree and Hercules'-club. And here I think a mix-up has occurred. It happened centuries ago and persists into the present day. I might have noticed *A. spinosa* several years before the Chief pointed it out if only I'd known what a neighbor was referring to when he mentioned the many Hercules'-clubs that he'd seen near our rural enclave. Birding, botanizing, driving to town, I kept an eye out for Hercules'-clubs. I knew exactly what to look for because I had met a representative of that particular species. It grew, and grows to this day, beside a nature trail that winds through a typical maritime forest—live oaks, laurel oaks, yaupons, loblolly pines—on one of North Carolina's barrier islands, Bogue Banks. The trunk of the tree, modest in every other respect, is clad in the most peculiar bark—bark that erupts all over with prominent bumps and humps. With knobs and warts, with buboes and carbuncles! They riveted my attention. Many were too big for me to encompass with thumb and middle finger, and all had been shaped by some arcane, arboreal geometry into protruding, many-sided forms. The tree's name fairly leaped from the field guide as soon as I got home. Its binomial, which sometimes begins with *X* instead of *Z*, is *Zanthoxylum clava-herculis*—the first part meaning "yellow wood" and the second "Hercules'-club." It belongs to the Rutaceae, the Rue family, which often smells strongly of citrus. And when I read about the tree's association with Hercules, I saw the Greek hero hard at one of his mythic labors: uprooting the tree, bashing its dangerously studded trunk hard against a giant's head, and strutting away then from the fallen enemy, whose skull had been deeply imprinted with dents in the shapes of pyramids and pentagons. Hercules'-club, yes, and once met, never forgotten. I would have recognized it instantly—had such a tree been growing anywhere in my neck of the woods.

No Hercules'-club nearby, but I did learn that it had gained John Lawson's attention back in 1709. Though the name by which he identified it, and spelled variously as "Pelletory" or "Pellitory," has fallen away like a shed leaf, he called it that because some of its properties resembled those of European pellitory, a plant used in the Old World to make toothpaste and to stimulate saliva.

> Pelletory grows on the Sand-Banks and Islands. It is used to cure the Tooth-ach, by putting a Piece of the Bark in the Mouth, which being very hot, draws a Rhume from the Mouth, and causes much Spittle. The *Indians* use it to make their Composition, which they give to their young Men and Boys when they are husquenaw'd. . . .

There was no confusion in Lawson's mind between Pelletory and Prickly-Ash. He plainly distinguished one species from the other, especially in regard to their medicinal applications. Just how *A. spinosa* and *Z. clava-herculis*, the A tree and the Z tree, came to be muddled together is not entirely clear, but a reasonable guess is possible. It might first behoove me, however, to turn off the alert sounded by a word like "husquenaw'd."

As Lawson trekked the Carolinas from the coast to the mountains and back again, he jotted copious notes not only on matters of natural history and geography but on the native inhabitants. Indian ways fascinated and sometimes appalled him. The practice of husquenawing certainly produced the latter effect; it was, he states primly, a "most abominable Custom." What he describes, however, is a typical scenario for the rite of passage often used by traditional people for marking a boy's transition to manhood and full membership in his tribe. In this case, the initiation took place in winter, and the boys involved were sequestered in what Lawson called a "House of Correction"—actually a large cabin set apart from the village. Feasting, drinking purgative beverages, they lived there in shivering darkness for five or six weeks before they were released, filled with new respect for their elders and knowing precisely what kind of strength they would need for their tribe's physically demanding way of life. It's the drinking and subsequent pandemonium that horrified our inquisitive but very proper Englishman. While the boys were isolated, Lawson wrote, they were given "Pellitory-Bark, and several intoxicating Plants, that make them as raving mad as ever were any People in the World; and you may hear them make the most dismal and hellish Cries, and Howlings, that ever human Creatures express'd."

Generation after generation, the husquenawers had always known what they were giving to the husquenawed. They knew their botany by heart and long experience, and knew precisely what plants would produce the desired effects. So it was never they who confused an aromatic, knobby tree with a scentless, spiny tree; never they who passed along such misinformation. Toothache tree, prickly ash, Hercules'-club—John Lawson can't be blamed for the mix-up either. Its source lies, I think, in the ease with which an untrained eye takes one thing for another because of passing similarities. One of them, in this case, is only bark deep: the spines. I did not see spines on the Hercules'-club growing beside the nature trail—but not because they weren't there. The sight of the weird bark, the lunatic geometry of its lumps and bumps, simply erased the possibility of seeing much else. It was the English naturalist Mark Catesby who clued me into the spines.

In 1731, Catesby, as keen as Lawson to observe and record the New World's wonders, published a drawing of the tree that he called "The Pellitory, or Tooth-ache Tree." This drawing, part of a natural history of the Southeast, shows a spray of the tree's long-stemmed compound leaves and a panicle of its greeny-yellow flowers, which serve as a botanical backdrop for his illustration of the ground dove. The accompanying commentary says that such doves "feed on the berries, which gives their flesh an aromatic flavor." And Catesby surely closed his eyes and sniffed with pleasure when he considered leaves that "smell like those of orange"; he must have pursed his lips and swallowed hard when he wrote of seeds, bark, and these same leaves tasting "very hot, and astringent." In fact he wrote so well and exactly about the tree that Linnaeus adopted his descriptive Latinate term, *Z. clava-herculis,* as its binomial.

And the spines? As boggled as I by that excrescent bark, Catesby noticed:

> The bark is white and very rough. The trunk and large limbs are in a singular manner thick set, with pyramidal-shaped protuberances,

> pointing from the tree; at the end of every one of which is a sharp thorn. These protuberances are of the same consistence with the bark . . . the largest being as big as walnuts. The smaller branches are beset with prickles only.

Thorns, prickles—not only was I so amazed that I didn't see beyond the protuberances, it's likely that the Hercules'-club I met was an older specimen that had long since lost the thorns once arming its warts and knobs, many of which were far bigger than walnuts.

The mere possession of thorns and prickles is enough all by itself to cause confusion in people to whom a tree is a tree, and if two separate trees both bear spikes and needles, then they must be the same kind of tree. Both the A tree and the Z tree are also pinnate, with dainty leaflets arranged in pairs along slender stems. And both bloom in large, airy, cloudlike clusters of many small, pale flowers. To the nonbotanical sensibility, it may not matter that A's leaves are not just pinnate but doubly so and that its flowers are perfect, while Z's blossoms come in male and female versions and occur on separate individuals. And where Z spreads out, with branches steadily diminishing to twigs, A produces at most a few spare, upward-reaching limbs. The two trees also differ in habitat, the Z tree clinging to the outermost margins of the Atlantic seaboard and the Gulf, the A tree rambling westward, occasionally crossing the Mississippi. (After all, it needs to be ready for quick plucking whenever the Devil decides to make his next excursion through the South.) And, though the A tree hogs all of the Z tree's common names—including angelica-tree, for Z's superficial resemblance to the European angelica once used for medicinal purposes—it lays exclusive claim to serve as the Devil's walking stick. That sobriquet is sometimes written out as "devils-walkingstick," as if pacing through the words at a fast clip might keep the Devil from springing into action at the mention of his name.

I waited for the young Devil's walking sticks at the edge of our yard to abandon their winter stillness. I watched each tree's one and only bud, the terminal bud set atop its slender, spiny trunk. Something was sheathed within. When the tight scales popped, what genie would burst free? The Chief watched me as I watched the buds and reiterated his question:"Do we really need those damned ugly things?" Of course we did, and I said so.

Midway through March, daffodils bloomed in the yard, jessamine in the woods. At month's end, the yellow-throated warblers returned and sang brightly from the tops of the pines. The Devil's walking sticks exhibited no signs of life. April came, and summer tanagers. Honeysuckle lent its fragrance to the warming air; sweet gums and tupelos unfurled their leaves. The bud atop each walking stick grew fat. One week into May, at the time that the multiflora roses in the hedgerows bloom and the migrating bobolinks pass through on their way to rice fields farther south, the walking sticks popped. They didn't just pop; they exploded.

Imagine fireworks. Imagine a skyrocket. When the buds flung themselves open, a stout, spiny branch shot out of each one, and shot up a foot into the air. And reddish leaves on long, thin, spiny stems spurted in sprays from each branch and arched over, falling toward earth in a glistening shower that never touched the ground but stayed high aloft. In mid-July, when this explosion was complete and the leaves had taken on a dark green color, a second stage was ignited, as dramatic as the first. Another stem thrust itself up, up, up above the leaves and detonated in a giant puff of tiny white flowers. There it stayed, like a crown or halo (or knob, perhaps, to ease the Devil's grip atop his stick). Nor did the show stop here. Green to pink to magenta, like sparks that flash and fade, the flower stems changed color, and white flowers transformed themselves to midnight purple berries.

Autumn equinox: time to gather berries and root bark, time to brew home remedies and effect a cure for what ailed me. No, neither a Cathartick nor an Emetick was called for. I certainly didn't feel cachexic. In this case, I suffered from nothing more than a critical inflammation of my curiosity. With long-handled spade to dig up roots and long-handled hawksbill shears to reach above spines and cut bunches of berries without impaling myself, it took me a short hour to collect materials. Another thirty minutes were required for stripping off berries, which were promptly covered with water and put on the stove to simmer. By noon I had made a good pint of Aralia-berry tea. Or not-so-good tea: it looked like tannin-stained creek water; its smell was that of earth and decaying leaves; it tasted bitter. But I drank it, softened its bite with liquid sweetener, and drank some more. Despite the warnings of John Lawson and others, nothing happened—no sweating, no purge at either end. I also washed and dried the roots. A concoction made later from their bark failed just as perfectly as berry tea to look, smell, and taste agreeable. The one virtue of berries and roots is that neither possesses spines. I learned well after these experiments with tea that *A. spinosa* had been summarily dismissed from its once honorable place in the official U.S. Pharmacopoeia at the beginning of the 1890s, more than a hundred years ago.

And I learned something else: if the Chief cuts down the Devil's walking sticks that thrive most cockily amid the decent vegetation in our yard, they'll show him what's what. They'll rise again. They'll sprout right up from underground runners and from the roots I left behind. They won't be extirpated. But if I find one day that they've really gone missing, how can I possibly blame the Chief? We'll both know that they've been taken up and away in a damnable but elegantly callused palm.

A HISTORY OF OBSOLESCENCE

Osage Orange

■

Now and in the past, Osage orange has been put to many uses, some expectable, some straight out of the blue. One of the latter was serving as Hedge, an obstacle around which Robbie Lankford made a significant end run. That story involves fire, chain saws, and Great-Uncle Leon, who walked on his ankles. It very much includes Rosalee Johnson. If she hadn't been there, I'd never have heard the tale. But more about all these things later, after Osage orange is given a proper introduction.

To begin with, the tree is known formally as *Maclura pomifera*. The species name translates as "pome bearing" and refers (inaccurately) to the large and peculiar fruit. The genus name honors William Maclure (1763–1840), a Scot who became an American citizen, served as a diplomat at the behest of Thomas Jefferson, surveyed much territory east of the Mississippi, made a great map, and became known as the "father of U.S. geology." He was also much involved in the utopian work-sharing community at New Harmony, Indiana, but that's another story. Suffice it to say that Osage orange stands by itself as the one and only member of the genus named for the industrious Scot.

But the tree is hardly without kin. It belongs to the Moraceae, the Mulberry family, which includes not only the mulberries, red and white, but the world's many figs. Mulberry trees, with slender, leafy branches cascading earthward like those of willows, gave delight to my childhood; they once lined the walks of the Virginia School for the Deaf and Blind, and in summer, when the students had departed, my friends and I swarmed to the ripening fruit, stood within the concealing branches, gorged ourselves silly, and staggered home with red-stained mouths and sticky hands—but that's also another story. All that really needs to be said is that mulberries and Osage oranges resemble each other in several ways. They both have orange-colored wood that weeps a milky juice when cut. Their compound fruits are not fleshy pomes like apples and pears but rather great bunches of single-seeded drupes that are intimately conjoined, one welded to the next. Though the mulberry is soft and purple-red and the Osage orange hard and heavy, big as a tennis ball and exactly the same shade of greeny yellow, both fruits are folded into convolutions as subtle and intricate as those of a brain, albeit a brain with hair, for the location of each drupe is signaled by a short, soft, whiskery strand. In the greenery department, Osage orange may be substituted for mulberry, the traditional fodder for silkworms; discerning no difference whatsoever, these caterpillars will devour the leaves of the former with a speed and gusto identical to that with which they have always attacked the latter.

Indeed, once upon a time in the later years of the nineteenth century, the federal government sponsored an effort at domestic silk production and, on request, sent out eggs that would hatch into silkworms. And so it was that in 1880 or thereabouts a shipment arrived in the hands of Miss Julia Rogers, who told the story twenty-five years later in her *Tree Book*. "Now gingham aprons were the prevailing fashion for little girls on the Iowa prairies," she wrote, "—princesses in fairy tales seemed to wear silks and satins with no particular care as to where they came from. Silkworms and Osage orange offered a combination, and suggested possibilities, which set our imaginations afire. . . . Not Solomon in all his glory was arrayed as we expected to be." She and another girl fed the hatchlings with lettuce and mulberry leaves, but when these little caterpillars grew into "lusty white worms so ghastly naked and dreadful to see, and so ravenous," nothing except Osage orange was available in a quantity great enough to satisfy the monstrous appetites. Risking injury from Osage orange thorns—and the tree is almost as spiny as a porcupine —the girls cut leafy twigs, and more twigs, and still more. Finally, the worms began to spin, and the bare twigs were decked with fine, fat cocoons. What next? Well, the cocoons hatched: ". . . the dead twigs blossomed with white moths whose beauty and tremulous motion passed description. We were lifted into a state of exaltation by the spectacle." Quickly, however, the girls' elevated spirits came crashing down. A neighbor informed them that breakage had ruined the silk, that the cocoons should have been scalded, their threads unwound, as soon as the spinning was done. Julia wrote, "Clouds and thick darkness shut out the day. We refused to be comforted." But to this day, despite the federal government's early encouragement, despite an abundance of Osage orange trees, caterpillar culture and silk making have never been successfully established in the United States.

The tree is sometimes called the bodark, a name that points to another of its former incarnations. "Bodark" is the pronunciation that resulted when Americans wrapped their tongues around the French phrase *bois d'arc*, which means "bow wood." And, indeed, the strong, flexible orange wood made splendid bows for Indian archers, particularly for the Osage, hunters of bison, whose homeland along the Osage River in western Missouri was part of the tree's original range. That range extended into Arkansas and the Red River valley of Texas. That was the region that the Osage came to call their own after they'd acquired horses from Europeans and noted the great interest of the French in trapping and trading furs. From the mid-1700s through the first two decades of the 1800s, the Osage successfully pursued trade, along with frequent raids on other people's horses, throughout the area. But starting in 1808, a series of treaties forced them to cede or sell mile on square mile of their lands in Missouri, Kansas, Arkansas, and Oklahoma. A population that numbered nearly six thousand in 1800 shrank in less than a century to a wretched one thousand five hundred. Their tale, however, follows the classic pattern of the underdog that at last wins out over the most formidable odds: the Osage Nation had retained the mineral rights to its former lands and eventually gathered in much wealth with the discovery of black gold. But this, too, is another story. Bodark's tale does not end on such an upbeat note. Though the rot-resistant wood of Osage orange is still used for such things as fence posts, it rarely serves archers. Bows these days are made mostly of high-tech plastics.

Technology has also bumped Osage orange aside in another of its traditional uses—dye making. Time was that someone who wanted a brilliant gold, a rosy orange, or a soft olive-yellow not given much to fading could gather the bark, simmer it to extract the color, and immerse skeins of wool in the resulting dye bath. The shade would depend on the color fixative with which the wool was treated before it was dyed. Some recipes for these colors often specified another, equally dependable dyestuff known as fustic, which comes from the tropical yellowwood tree (*Chlorophora tinctoria*) that grows in Cuba, Jamaica, and Central America. One of its nicknames, dyers' mulberry, makes clear its close relationship to Osage orange; both are among the Moraceae. Early on, fustic was a closely guarded treasure; a British navigation statute of 1661 listed it as one of the colonial products that could be shipped from America only to other places under British rule. Into the twentieth century, fustic was well regarded as a source of every shade from drab tan to gold and orange. But in the days of the First World War, it may not have been easy to import a tropical tree—or to import it quickly enough. To the rescue came Osage orange, which provided the khaki color for countless yards of woolen goods that were cut and stitched into soldiers' uniforms. But modern times have seen the natural tans, yellows, and golds of both fustic and Osage orange displaced by upstart colors brewed in chemical vats.

I'm sneaking up on Robbie Lankford and his end run. But there's one more short byway to explore before Hedge takes over, as Hedge is wont to do (it deserves the capital letter for its tough and prickly presence, as well as the fact that an expletive—damned Hedge! —often accompanies mere mention of the stuff). Some people refer to Osage orange as the "horse apple." The reason usually given for the origin of this epithet is that the tree's large, green, applelike fruit, which human beings find unpalatable, is readily eaten by horses and other livestock. Really? Though I have not consulted any horses on the matter, another reason seems more likely: the nickname surely refers not to what goes in one end of the horse but to the size and excellent roundness of what comes out the other.

And here he comes—Robbie Lankford, racing around the end of that thorny Hedge on his way to meet Rosalee Johnson. He'll outfox all the disapproving Johnsons, including Rosalee's brother Leon, and he'll marry her. It's their grandson, Bob Marlin, who tells me their story and, more important, gives details of his own close acquaintance with Hedge.

This particular Hedge was one of many such living fences planted throughout the Midwest between the 1860s and the late 1930s. Miss Julia Rogers, whose adventure with silkworms has been related, made these remarks about Hedge in her 1905 *Tree Book*:

> The Osage orange hedge marked one period in the pioneer's work of taming the wilds of the Middle West. Farms had to be enclosed. Board fences were too costly, and were continually needing repairs. Fencing with wire was new and ineffectual, for barbed wire had not yet come into use; so hedges were planted far and wide. The nurserymen reaped a harvest, for this tree grows from cuttings of root or branch. All that is needed is to hack a tree to bits and put them in the ground; each fragment takes root and sends up a flourishing shoot.

These drought-hardy, sun-loving trees were set in as windbreaks, too, and the federal government not only encouraged but supervised their installation on midwestern farms to keep the soil from blowing all to hell and gone in the dust bowl days. Hedge was easy to put in because of its habit of forming a coppice—a dense thicket, a small forest—from suckers sent skyward out of established roots. It's been estimated that by the time Hedge went out of fashion, some 96,000 miles of Osage orange had been planted—nearly enough to circle the earth four times around the equator.

The living fences with which my friend Bob became well acquainted some forty years ago were simply called Hedge—but without the addition of expletives, for the family were Baptists.

When Great-Uncle Leon, who walked on his ankles, was exceedingly provoked, he sometimes muttered, "Mad Dog spelled backwards." The ankle-walking was the result of some misfortune in infancy, polio perhaps, that had turned both feet under; for the rest of his life he hobbled around without a cane, and the soles of his shoes never touched the ground. Great-Uncle Leon farmed land that his father had bought from the Lankfords. It was situated in the home territory of Osage orange, near the now vanished town of Mayview, Missouri, some fifty miles east of Kansas City. It was there that Bob first encountered Hedge. And what a mighty Hedge it was—at least forty feet high and some thirty feet deep—separating the Johnsons from the Lankfords (whom the Johnsons considered lazy because they'd done nothing except come west to Missouri in the early 1800s, take up a section or two of farmland, and then sell off pieces and parcels). It was also an enthusiastic Hedge that had to be trimmed back at least once a year. To trim it, the Johnson boys used two hay wagons, one on either side of that Hedge, and a double-handled crosscut saw that they worked back and forth till the trees were cut down to the shoulder height of a man standing on a wagon. Though the Hedge lay on their mutual property line, with most of its thorny, thickety tangle on the Lankford side, the Lankfords never helped but watched closely to see that the Johnsons did not try pushing the line over onto land that wasn't theirs. Great-Uncle Leon, who was in his sixties at the time that Bob met Hedge, didn't help either; he must have had a painful encounter with its bristling spines on some earlier occasion, for Bob never saw him get closer than six feet to a living hedge tree.

It was not on this monumental Hedge, however, that Bob was first put to work, but on a lesser Hedge, no more than twenty feet high and only fifteen deep, on a Lankford farm the other side of Mayview. The job: to take the whole thing out, to banish it forever from the face of the land. The adults in charge, taking on Bob and his second-cousin Billy as apprentices, used a method that Great-Uncle Leon regarded as sacrilege, though not for reasons of religion.

As he saw it, the Lankfords were controverting time-honored ways of handling Hedge by using newfangled stuff to burn it. Gasoline, heaven forbid!

"There were gasoline fumes in the air," Bob says, "and the wheat stubble was black as we approached the smoking remains of the hedgerow. Billy's dad had burned it green and standing. Burning removed the thorns and some of the small branches. I can still see those black trunks, some the size of my leg, some smaller. Since I was only thirteen, they wouldn't let me use the chain saw unless the trunks were already down, and then only for a cut or two."

Bob's task was to rescue both firewood and potential fence posts from the char. Wearing a thick pair of mule-hide gloves, using a brush hook and a hay hook, he dragged post-sized logs from the smoldering stubble. When they'd cooled a little, they were stacked in vertical piles like corn shocks. Firewood was taken on flatbed wagon to the woodyard behind the house.

"And Lankfords were out there making fence posts in the hot sun and black smoke—hundred-degree weather, when not even a Baptist in his right mind would be out in the fields. We followed up with a dip in the stock pond. These cousins weren't Baptists, but they were some kind of Christians; so, there was no skinny dipping. But we did get nearly naked and wet in that green water just the same, along with the ducks and frogs."

Everyone but Great-Uncle Leon eventually left the farm, and he was old then, though still getting about on his ankles, if a bit more slowly. Without supervision, Hedge went wild, sending up new shoots, creating great thickets, running over property lines without the slightest regard for Johnsons or Lankfords.

Nobody plants Hedge these days. To be sure, remnants of the great Osage orange barricades still stand, as thick and thorny as ever. The current thrust, however, is toward eradication. Hedge has gone the way of bow wood, dye making, and fodder for silkworms. Like them, it fills no contemporary needs.

But try telling that to an Osage orange. It's only the tree's human uses that have suffered obsolescence. The tree itself thrives. Now growing far beyond its native range, it is perfectly at home in both the East and the Northwest. Each fall, several Osage orange trees drop copious quantities of fruit upon my hilly street in a small Shenandoah Valley town. One nearby side street is even named Osage Place for the venerable, five-trunked tree with many branches overhanging the road. The heaps of fallen fruit glow—neon chartreuse on black macadam; eventually, traffic smashes the mess. On the other side of town, my nephew must pick up bushel on bushel of greeny-yellow balls before he can mow the autumn grass. And my elder daughter may be courting such a fate; when she visited two weeks ago, she collected several large, firm fruits, newly shed by the tree on Osage Place, and took them, along with a box turtle, to her home in Wisconsin. That turtle, of course, is another story.

Let this one end by giving praise to Osage orange's rooted stubbornness. In its own vital terms, the species will not—cannot—become obsolete. Obsolescence is a human conceit anyhow, and one that's not available to treekind. It's possible, yes, that our kind might manage, by accident or intent, to pull off a massive but not entirely unthinkable feat of tree removal. But in that case, we'd be out of a tree, probably our own.

A GOOD TRUCK-FIXIN' TREE

The Catalpas

■

"It's a good truck-fixin' tree, I'll tell you that. At least, for a shade-tree mechanic."

So says my elder son, Pete. But he's actually been using that big old catalpa tree for something considerably more dramatic. He's attempting nothing less than truck resurrection. The truck is a red 1986 half-ton GMC pickup, which in its active days bore a proud vanity plate: MY TRASH. It met disaster two years ago on a dark, rainy evening, the kind that makes roads slicker than a greased pig, with headlights bouncing off the oiled surface with a harsh white glare. As Pete was driving home from work, a van cut sharply in front of him from the left-hand lane. To avoid collision, he put the pickup on the shoulder, but when he tried to stop, the brakes locked and the truck skidded into a metal utility pole. He wasn't hurt, but his trash suffered massive damage—fender wrapped around a door, bent frame, front prop shaft through the aux box, radiator pushed into the fan blade, motor mounts twisted. That truck red as heart's blood was totaled.

The wreck was towed to my driveway, where it languished for an inconvenient month or two (I wanted my off-street parking back). Then, one of Pete's friends with access to a roll-back picked it up, carted it off, and set it down a couple of yards from the catalpa tree.

I first met the tree in winter, raising bare black branches fringed with many seedpods into a cold, bleached-white sky. It stands in a yard just off the intersection of Goose Creek Road with US 250. The Blue Ridge mountains rise to the east, and in the west, Great North Mountain stretches for miles like a hazy grey-purple wall. The yard is small and neat, except for the dozen or so chickens pecking through the grass and a collection of a dozen or so old cars and trucks parked at its edges. Goose Creek meanders along the back line. Though it's shallow and not more than three feet wide, mallards regularly swim and dabble there, and Pete has seen an occasional beaver.

When the tree is in leaf, it shades not only chickens, trucks, and my son the shade-tree mechanic, but also a light green frame ranch house of modest proportions. Or, it looks modest from the outside. But walk in the back door, which is located a mere six feet from the catalpa tree's stout trunk, and you'll enter a sizeable room that holds a regulation pool table and a bar ten feet long. Often as not some hirsute characters with long ponytails and many tattoos will be there laughing, telling jokes, drinking beer or bourbon, and just generally hanging out in a haven free from women—unless, of course, the women have been invited. These are the Scavengers, members of a motorcycle club dedicated to the riding and loving preservation of Harley Davidsons; the light green ranch house is their home away from home. Despite the club's name and vulturine emblem, despite their fierce and bristling mien, Pete and his cronies are not to be equated with gangs like the Hell's Angels or the Pagans. Reveling in their appearance—something that would strike envy in a Mongol horde—the Scavengers turn out en masse, showing off, flaunting their colors, to vote, give blood, and distribute toys to children.

As Pete's mother, I seem to be exempt from the interdiction against uninvited women. The club members also know that I don't go to the ranch house on Goose Creek Road for anything so frivolous as hanging out. I go there to visit the catalpa tree.

And it's a fine tree indeed, an excellent specimen of *Catalpa bignonioides*, the southern catalpa. It reaches upward for a full 60 feet, exactly matching Virginia's current record for the species. To arrive at this figure, Pete's totaled red pickup comes in handy. Looking at a photo that includes both tree and truck, I calculate the tree's height in multiples of the known height of the cab. By the same cab yardstick, its crown spread may be estimated at 45 feet, much less than the state's 77-foot record. There's a reason that this tree falls short: many of the lower branches, some nearly a foot in diameter, have been neatly cut away so that they do not thrust against the house. But the branch that figures in truck fixin' has never been trimmed; it juts out horizontally for several feet, then turns skyward at a right angle, like an arm flexed to show off the biceps—but more about that later. As for the trunk's circumference, I don't need to make an estimate. I simply recruit a Scavenger with a steel measuring tape to get a reading at breast height, or four and a half feet off the ground, which is the canonical location for ascertaining trunk size. (Diameter at breast height—DBH—is what the botanists call this vital measurement. No one I ask seems to know why, nor whose breast figured in the original formula.) The tape says that this tree's girth is 7 feet 11.5 inches, for a DBH of slightly more than 2½ feet. Trifling compared to the state record of 18 feet 5 inches, but it's more far more than a single tree-hugger can possibly handle.

No catalpas grew near Goose Creek Road nor anywhere at all in the Shenandoah valley three hundred years ago, when colonists began to trek westward over the Blue Ridge and settle down in significant numbers. *C. bignonioides,* the southern catalpa, was then a species found only in the rich, moist bottomlands of the deep South—from South Carolina through Georgia and northwestern Florida into Alabama and Mississippi. Nor had its cousin, *C. speciosa*—the western or hardy catalpa, escaped the confines of its original range from southern Illinois to northeastern Arkansas. Both species have since made themselves widely and comfortably at home in the eastern United States. Just who sent the western catalpa out into the greater world is not known, but the initial responsibility for the southern catalpa's spread can be most precisely assigned to an Englishman named Mark Catesby (1682–1749).

"This tree," he wrote in his *Natural History of Carolina, Florida, and the Bahama Islands,* "was unknown to the inhabited parts of Carolina till I brought the seeds from the remoter parts of the country. And though the inhabitants are little curious in gardening, yet the uncommon beauty of the tree has induced them to propagate it; and it has become an ornament to many of their gardens, and will probably be the same to ours in England."

By "inhabitants," he meant only the European colonists. And he did indeed take catalpas to England, where they flourished right merrily in the botanical collection of one Mr. Bacon of Hoxton, then a rural town that has long since been encompassed by greater London. (He also took home some poison ivy plants, which he raised successfully and later exhibited to fellow members of the Royal Society.)

Catesby's adventures in Carolina and other parts of the South were inspired by a passionate curiosity about the New World. And he realized his goal, clearly stated on the title page of the *Natural History,* to draw "the Figures of Birds, Beasts, Fishes, Serpents, Insects, and Plants: Particularly the Forest-Trees, Shrubs, and Other Plants, Not Hitherto Described or Very Incorrectly Figured by Authors." The *Natural History* appeared slowly between 1731 and 1743 as he found the money to have another group of its 220 plates engraved and colored. Each plate is accompanied by descriptive notes and personal comments. Fully half of them portray both a bird and a plant. The leaf of the southern catalpa—"shaped like that of the lilac, but much larger"—is shown with the bird that Catesby called the "Bastard Baltimore," an orchard oriole, whose soot-and-rust coloring he evidently scorned as an imperfect imitation of the Baltimore oriole's bright flames. What marvelous vigor inhabits this plate and the others! Though the drawing is often primitive, with the birds posed in the most graceless postures, each still brims with its maker's exuberant wonder at all he beheld. His memory is honored today in the scientific nomenclature: the bullfrog, *Rana catesbiana*, bears his name.

The catalpa received its own name, the name still in use, long before any European set foot on these shores. *Kutuhlpa*—that's what it was called by the Creek Indians, who lived originally in an area destined to become part of Georgia and Alabama. (Later, the Creeks would accompany the Cherokees on the Trail of Tears into Oklahoma Territory.) Their word means "head with wings" and describes the frilly, outspread lobes of the flower petals. In May, both southern and western versions of the tree are a glory, decked out with lightly fragrant, creamy white blossoms—dotted with purple, brushed lightly with gold—that hang in flouncelike panicles amid the leaves. And the leaves are huge and heart-shaped, casting a definitive shade on all who work, or lollygag, below. When they're shed each fall, the trees are still not unadorned; seed-bearing capsules dangle at the tip of every branchlet and twig. These pods, though no bigger around than my little finger, grow into slender spindles from six to twenty inches long. Green at first, they darken at the onset of colder weather and split, releasing flat, winged, silvery or light brown seeds that seem no more substantial than snowflakes. The empty pods, still clinging to the trees, rustle and clack on the passing wind.

As with other trees native to North America—persimmon, sassafras, and tupelo, for three—the nomenclators knew a good thing when they heard it and selected catalpa's native name as the designation for its genus. *Speciosa,* the western catalpa's species name (its trivial name, in the lingo of taxonomists), means "showy" and expresses admiration for those great clusters of gorgeous blossoms. *Bignonioides,* the formal name for the truck-fixin' tree and all other southern catalpas, translates as "bignonia-like," a term that again refers to the flowers. Mark Catesby described these as "tubulous"; they're made in the shape of a trumpet with a boldly flared lip, a characteristic that catalpas share with the orange-red trumpet-creeper (*Campsis radicans*), the golden-throated cross-vine (*Bignonia capreolata*), and many other members of their family, the Bignoniaceae. (The family 's name honors the Abbé Bignon, librarian to France's Louis XIV, who reigned from 1643 to 1715. Because his name is French, its *g* and *n* are not separately pronounced but rather softened into a n-yo sound. Thus, despite the dictates of Webster, the species name of the southern catalpa is said this way: *Been-yon-ee-OI-dees.*)

Only seven species of catalpa exist in the whole world: two in the United States, one in the West Indies, and the rest in the Far

East. Like persimmon, pawpaw, tulip tree, and a surprising lot of other trees found in the Americas, the genus has close kin in Asia but no representatives native to Europe. The vines, shrubs, and trees of the Bignoniaceae family are found most abundantly in tropical regions of the New World. The two catalpas of the U.S. flout family convention; they're upstarts, tolerant alike of summer scorchers and winter snows.

And our two catalpas resemble each other almost as closely as identical twins. They even share a pest: *Ceratomia catalpae*, the spike-tailed caterpillar of the catalpa sphinx moth. The creature is a cousin of the tomato hornworm, and equally voracious. The larval appetite is indeed so great that a heavily infested tree may have its leaves completely stripped. But that doesn't happen often, for this juicy caterpillar attracts its own pests. Not only is it highly vulnerable to attack by a parasitic wasp, it's also eagerly grabbed up by fishermen. With a twig or a wooden match, they turn it inside out (so that a fish need not work very hard to get at the good part?) before it's impaled on a hook as bait.

How to tell one species of catalpa from the other? At the easiest, it's a matter of catching them in bloom. The panicles of the southern species bear their flowers in denser array; the blossoms are also somewhat smaller but more extravagantly dappled with purple than those of their western cousin. The leaves of the western catalpa are odorless when crushed, while those of the southern tree give off a faintly rank smell. Subtle differences in other characteristics, such as the shapes of leaves and seed capsules, serve also to distinguish them. The somewhat hardier western catalpa once outranked its southern counterpart (though not by much) as a provider of valuable timber —fence rails and posts, railroad ties, telegraph poles. A virtue of both species is their habit of fast growth; a sapling would attain a size suitable for posts in only six years; a telegraph pole, twenty-five. At the turn of the century, plantations of western catalpas took up considerable acreage in Kansas and other parts of the Midwest. Such enterprise has been bumped aside, however, by the ubiquity of other materials, from treated pine to metal and plastics. No matter; both catalpas are esteemed as ornamentals for their cool green shade and the May-time beauty of their flowers—not to mention a certain reliable sturdiness when it comes to fixin' trucks.

As it happens, the southern catalpa by the back door of the modest light green ranch house on Goose Creek Road is good for more than truck repair. It's a fund-raising tree and a tree for dressing out white-tailed deer. It could also be smoked.

If the Scavengers were kids, that's what they'd do—smoke the catalpa beans, as common parlance has nicknamed those long, dark, slender seedpods. But most of the Scavengers don't know the source of the pods as a catalpa, nor would they recognize it by another of its common names, Indian bean. They call it the "cigar tree." I've heard people elsewhere referring to the pods themselves as "monkey cigars." And my husband, the Chief, is well versed in smoking them. In his elementary school years, when he was barely out of the yard-ape stage of his life, he and his cronies in rural North Carolina all turned into cigar monkeys—that is, when they weren't smoking rabbit tobacco. They'd break off both pod ends, light up, and puff themselves into delirious dizziness. As for me, at that tender age I was certainly familiar with catalpa beans but never once attempted to smoke them (but that may have been because no one ever suggested such an experiment). My childhood experience consisted of stomping through fallen beans—crunch, crunch—and kicking them off the sidewalk in the small town of my growing up. (These days I've become aware that some people hate catalpas—that's right, *hate* them—for the mess they make. Late winter or early spring, in certain parts of town, the seedpods fall and litter the sidewalks; glaring, tightening their lips against unladylike imprecations, women sweep them briskly into the street.)

As for raising funds, dressing out deer, and repairing trucks, all these activities depend, quite literally, on the chain that's wrapped twice around the muscular branch located fourteen feet up the trunk of the tree. And the chain is a heavy-duty log chain with 3/8-inch links. It's fastened in place with grab hooks to keep it unbudgeably in place.

A whitetail, of course, is hoisted aloft by a rope looped through the chain and tied to its hind legs. For the fund-raising gala, an affair that lasted all night, the chain held a 650 cc. rice grinder (as the Scavengers and others of their kind impolitely refer to Japanese-made motorcycles).Anyone interested in working off aggressions—or simply in yelling and carrying on—could spend a buck and buy the chance to whack at it three times with an eight-pound sledge-hammer. When the handle of that hammer broke, someone supplied a ten-pound model. Enough money came in to complete the building of the addition that houses the club's pool table and bar.

I have to ask Pete just how the chain on the catalpa branch figures in truck fixin'. He pencils a sketch on a paper napkin and explains.

Ah, I see. The truck sits in the clubhouse driveway, the tree in the yard. A four-by-four-inch board, with a strap attached to each end, extends through the windows of the cab. A chain jack, anchored by the chain on the mighty catalpa branch, is pulled over and hitched to the straps on the four-by-four. Then the shade-tree mechanic can winch the truck off the ground and move it into whatever fixin' position he deems most suitable.

There's not much left to fix on the red 1986 half-ton GMC pickup—rebuild the carburetor, blow out the brake lines, tighten up a few things here and there. That truck's gonna rise again. Its next vanity plate ought to read LAZARUS.

LIFE STORIES

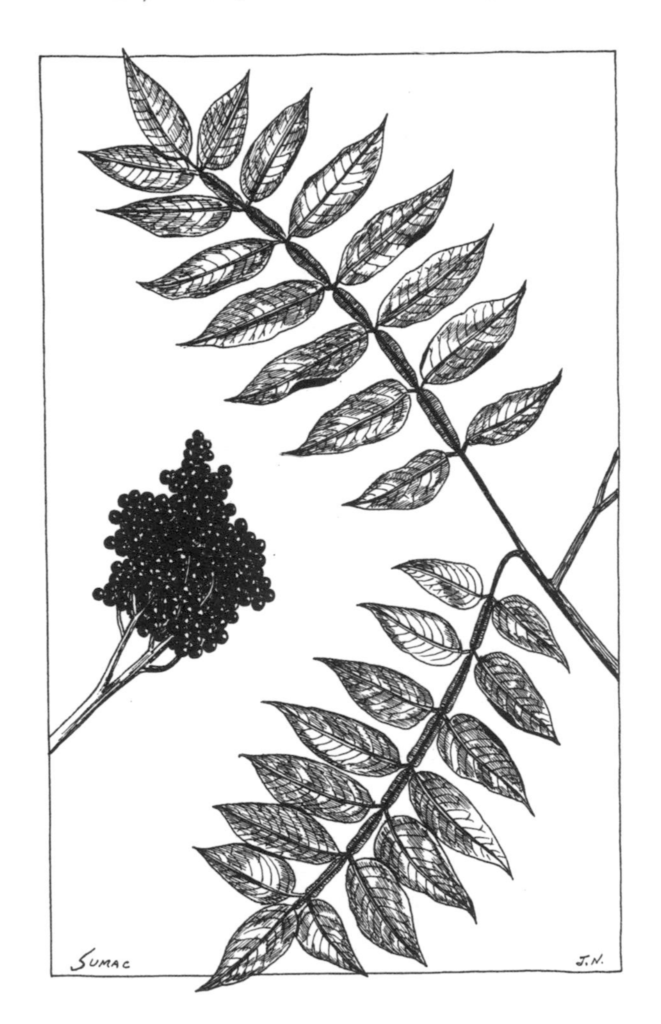

The Sumacs

■

My nature is subdued
To what it works in, like the dyer's hand.

— William Shakespeare, Sonnet 111

Hands working in dye made from sumac can turn grey. Not charcoal grey nor ash, but rather the dark, dirty grey of a slag heap. And they stay that way for a full week, with traces that persist far longer around and under the fingernails. Only passing time and the shedding of skin cells are able to effect a cure. But, oh, the rewards! Wool simmered in a dye bath made from sumac berries and ferrous sulfate will dry to other shades of grey—grey warm and sleek as a catbird's feathers, grey cool and blue as winter dusk, grey darkly thunderous as an oncoming storm.

Twenty-four years ago I dyed skein after skein of four-ply wool with sumac and other vegetable matter and used the wool to make a hooked rug, a runner two and a half feet wide by seven feet long, that now covers the floor along one side of the very large table that is my desk. Today, the sumac's grey holds fast. Not sunlight, not decades of foot traffic, not even the puddles made by several unhousebroken puppies have stolen a whit of that faithful grey. The dyes made from goldenrod, loose-leaf Salada tea, and the lichen mixture known as cudbear have lasted nearly as well, but the others —dandelion, barberry, fresh and dried mullein, seaweed—suffered fading almost from the first moment that they were exposed to light. But they have not disappeared, nor is the wool itself unduly worn. The colors have simply become softer and friendlier. And the story that they illustrate in fourteen panels is still being told from beginning to end—except that the story has not truly reached its end and, with fortune and forethought, never will.

73
J.N.

I found the story in *The Book of Signs* by Rudolf Koch, who won great acclaim in his native Germany for his elegant type designs and bookbinding. Religious crosses, runic alphabets, maker's marks and marks of ownership, the artfully stylized squiggles of alchemy and astrology—he had noted a grand variety of ancient and medieval symbols and, for this book, re-created 493 of them in his clean, calligraphic hand. My eye was caught by a set of related symbols from the Middle Ages—a life story. And the tale it tells in the plainest geometric lines is that of

a man
            and a woman
who join in marriage.
            The woman becomes pregnant
and bears a child.
            Soon, the family consists of man, wife,
            and two children.
And the man has a friend,
            but they come to blows,
and the man dies,
            leaving his widow with the two children.
When one of the children dies,
            the mother lives on with the surviving child
until she, too, dies,
            leaving the child, who contains the promise
            of future generations.

The figures were worked in wool dyed with cudbear, one skein for each of three successive dye baths: deep purple, maroon, medium pink. Sumac grey, the color of a thunderhead, outlined the squares in which they were set.

And just as human lives rise and fall and rise again on a great rolling wheel, so do the four seasons. Their symbols, in a dark green made from goldenrod fixed with copperas, accent the rug's outermost corners: spring's round seed that has just begun to sprout, summer's invitingly open cup, fall's overturned cup pouring forth a bountiful harvest, and winter's closed-in world, with snow falling in immense flakes from a sky that nearly touches earth. Two more panels, cudbear again and worked immediately after the fourteen giving the family's story, show my initials and the year of making. I lived in Connecticut then, six miles from Long Island Sound.

The other stories in the rug are not so outspoken. They hide in the colors. They whisper of many things. One set of tales is concerned with collecting the natural materials from which the dyes were made. These are all part of my own life story. I can read the rug as I would read a journal or a diary. Take seaweed:

A book of recipes informed me that it could yield shades from dark yellow-green to grey and, better still, that they would be fast. The next move, of course, was to pack the car with buckets and English springer spaniel and head for a local beach, the western end of which featured a breakwater of seaweed-covered boulders. In summer's long, hot tourist weather, the dog would not have been allowed to set paws on the sacred sand, but we made our excursion at the end of October. Though the sun shone warm in a bright blue sky, the beach was deserted. And, oh, the ebbing tide had exposed a lush crop of slippery, yellowish weed—rockweed, with air bladders that could be squeezed or stomped to make a satisfying *pop!* Spaniel dived into the salty water of the sound and paddled happily like the water dog she was. Woman clambered over skiddy rocks, slipped, laughed, tried again, and eventually gathered three buckets of weed. Later, as it simmered in a canning kettle so that color could be extracted for dye, the children, just home from school, sniffed the air, held their noses, and said, "What are you cooking *now?*" and "If that's dinner, I'm out of here."

Another set of the rug's stories deals with dyers and making vegetable dyes, and with the lovely, resonant, though sometimes peculiar language that's used to describe the stuffs and techniques of an ancient craft. *Mordant* refers to a chemical—be it alum, chrome, blue vitriol, copperas, tin, or something else—that fixes a dye in yarn or fabric. The word means "biting"; I see the fixative clamping down so hard that color cannot escape its grip. *Copperas* is a fourteenth century term for a hydrated ferrous sulfate. It can be used especially *to sadden,* or dull, a color, while *blooming* adds brightness. And the names of the dyes, the plant names, roll and spit and thud: sumac, goldenrod, marigold, quercitron, sassafras, indigo, catechu, fustic, madder, weld, woad, cudbear.

It's time, yes, to explain cudbear. Unlike the other names on this skim-the-surface list, the term does not designate both a plant and a dye, but only the latter. It comes from Cuthbert Gordon, the Scottish merchant who patented this mixture of lichens in 1758; he chose to honor his mother (as well as himself) by giving it her maiden name, which, not coincidentally, was Cuthbert. The name of the dye is sometimes written out as "cut bear," a spelling that may reflect a Scottish attempt to frenchify pronunciation of the family name. Cudbear is the one dyestuff I mail-ordered from a commercial source; the others came from places close to home—fields, woods, yard, or, in the case of loose-leaf tea, the supermarket. But stories like these are particular to an individual or a craft and, like the tales of my collecting expeditions, only fragments of a whole.

But another, greater set of stories also reposes in the rug, the stories of the plants that yielded the dyes—dark green and rosy tan, purple, pink, and stormy grey, the several yellows. And every story here is a life story with the same grand impetus, the same urgency as the one told by the medieval symbols. One narrates the life and times of barberry—every barberry, another of all the goldenrods, and yet another of the three distinct lichens that grant their imperial purple to Cuthbert Gordon's patented dye. The story to which I'd like to put words is the sumac's tale, of course, for sumac is the only arborescent plant in the rug's garden of colors, the only one able to grow into a tree.

Time out of mind, sumac has been used as a source of dye. And not only that, but members of its large genus, *Rhus,* have variously provided tannin for curing leather, sap for the manufacture of spirit varnish, leaves to be mixed with tobacco and smoked, and berries that make a most refreshing lemonade-like drink. Some *Rhus* species are to be avoided because their resin or volatile oil is poisonous, causing human skin to erupt in excruciatingly itchy blisters. One *Rhus* species, the North American *R. glabra,* or smooth sumac, is reported to have given at least one man the power to fly.

The name for the genus has been in continuous use for several thousand years. *Rhous* is the ancient Greek word used for the tannin-rich sumac of southern Europe and the Near East, *R. coriaria,* or "leatherlike sumac" because of its tough leaves. And *rhous* comes from *rhein,* a Greek verb meaning "to flow," perhaps because the plant was thought to cure watery, gushing ailments like dysentery and diarrhea (the latter a word that also incorporates *rhein* and means "flow through"). And long after classical times, sumac retained its reputation as a sure remedy for stanching the runs and other unwanted bodily effusions. In his 1597 *Herbal,* the English botanist John Gerard listed many such medicinal uses of the plant: "The leaves of Sumach boiled in wine and drunken, do stop . . . the inordinate course of women's sicknesses, and all other inordinate issues of blood. The seed of Sumach eaten in sauces with meat, stoppeth all manner of fluxes of the belly, the bloudy flix, and all other issues, especially the white issues of women." The last refers to leukorrhea (*rhein* again—"white flow"). As for the word *sumac,* its immediate ancestor is the Arabic *summaq,* which names the same plant that the Greeks called *rhous;* it was imported into English sometime during the crusading 1300s, when Europeans confronted Muslims in the Holy Land.

The sumacs in their many guises belong to the Cashew family, the Anacardiaceae, found in tropical and temperate climates around the world. The relatives of the eponymous cashew tree include, in some seventy genera and six hundred species, the mango, a pistachio native to Texas, and the truly ornamental American smoketree (*Cotinus obovatus*), which is covered in spring by a fine cloud of pink blossoms, in fall by handsome red and orange leaves. (Alas, the Texas pistachio produces no nuts, and the American smoketree is far less commonly used in landscape plantings than its showy European cousin.) The whole family is notable for its poisons; not just *Rhus* but many other genera boast species that are exceedingly unfriendly to humankind. The cashew is one of them; the husk of the West Indian nut (*Anacardium occidentalis*) contains a substance so caustic that it was used by indigenous people to scarify their skins with traditional tribal designs.

Around the globe, *Rhus* species demonstrate the family's inclination toward toxicity. In North America, there's no escaping the truly obnoxious members of the genus: *R. vernix*, the "varnish" sumac; *R. radicans*, the "rooted" kind; *R. quercifolium*, the "oak-leaved" species; and *R. diversilobum*, the species that's "variable leaved." In plainer English, these are none other than poison sumac, poison ivy, poison oak, and Western poison oak. Because of their toxicity, some botanists would place them in a genus of their own, *Toxicodendron*, or "Poison tree," but they seem to possess no distinctive quirks of seed, leaf, and other features that would set them apart from other members of the genus *Rhus*. But there is an easy way for those of us who aren't botanists but simply amuse ourselves going up and down hill to tell the harmless sumacs from those that will surely cause grief:

> Berries white, take fright.
> Berries red, go ahead.

At least, that method will do when the plants bear fruit. In other seasons, it's wise to know the shapes of the different leaves—or just to stay clear.

North America's poison ivy and poison sumac: in human terms, these plants seem good for nothing at all—nothing, that is, but causing misery. Among the many dire effects of contact with poison sumac are these, as recorded in 1892 by the American pharmacognosist Charles Millspaugh in his then definitive work *Medicinal Plants*:

"Sadness and gloomy forebodings; vertigo; dull, heavy headache; profuse watery stools; . . . trembling of the limbs; . . . almost all forms of skin trouble, from simple redness and burning, to vesicles, cracks, pustules and complete destruction. . . ." (We shall hear more from Dr. Millspaugh shortly.) In *Domestic Manners of the Americans,* Mrs. Frances Trollope commented on yet another disadvantage of these two unfriendly species of *Rhus,* which she called by names common in her day—a disadvantage as potentially devastating to human flesh as any mentioned by Dr. Millspaugh. With friends, she visited the great falls of the Potomac River and later recorded her awe at the cataract's twisting, roaring, dizzying violence as it plunged over the rocks. There, amid the crash and thunder of falling water, she also noted that "the slenderest, loveliest shrubs, peep forth from among these hideous rocks, like children smiling in the midst of danger. As we stood looking at this tremendous scene, one of our friends made us remark, that the poison alder and the poison vine, threw their graceful but perfidious branches, over every rock, and assured us also that innumerable tribes of snakes found their dark dwelling among them."

But can it be that stuff so hospitable to serpents and so noxious to humankind does not possess one single virtue? Well, according to a French report published in 1557–58, the Indians of the St. Lawrence employed poison ivy as a defense in war; it was placed with other baneful vegetation on stacked wood greased with fish oil and set afire if enemies came barreling in with the wind against them. And the milky sap of poison sumac has been made into a black varnish quite as fine, hard, and lustrous as that made time out of mind from the equally toxic oriental species, *R. vernicifera,* the varnish-bearing sumac, which is not only tapped in the wild but grown in plantations. But it's the herbalists who've had a field day. Operating on a notion similar to the one that states, "If it tastes bad, it's got to be good for you," they have diligently investigated the possibilities of transforming the toxic juices of these vines and shrubs into healing remedies. Since at least the late 1700s, potions brewed from the juice or poultices made from crushed leaves were rumored to ease paralysis and act as a stimulant. Throughout the 1800s, despite significant skepticism on the part of many doctors, much effort was put into attempts to extract useful medicaments from these pernicious weeds. The resulting nostrums, swallowed or applied externally, were touted as working wonders on a Pandora's box of diseases, from consumption, chronic asthma, rheumatism, and palsy to ringworm and chicken pox. None, however, was claimed to cure the blisters and raging itch caused by the toxic species of *Rhus*, although the antidotes for such poisoning are, as Dr. Millspaugh wrote, almost as many as those for the bite of a rattlesnake. No one has yet turned this particular sow's ear into a silk purse, though even today, on the brink of the twenty-first century, people keep trying. A small North Carolina company that deals with wild botanicals will pay as much as two dollars a pound for all the poison sumac, ivy, and oak that foragers can lug in; its factory processes the plants and sends the resulting extracts to Germany, where they become principal ingredients in an ointment said to palliate the aches and pains of arthritis. According to the company, its suppliers are among the lucky souls immune to the poisonous oils in these vines and shrubs. Not so. No one is entirely safe, for each exposure increases sensitivity; those who have escaped ill effects for decades may find themselves suddenly inflamed, both inside and out.

Sensitization may have accounted for Dr. Millspaugh's most astonishing—indeed, otherworldly—adventure with sumac. The sumac in this case is not a noxious species, but rather *R. glabra*, the smooth sumac; its bark, especially that of the root, was used in the nineteenth century to treat a hodgepodge of diseases, including ulcers and gonorrhea, and juice from its roots was said to make warts disappear. Scouting out such medicinal plants, Dr. Millspaugh spent considerable time making personal expeditions through the underbrush. And there, in the course of his botanizing, he was often exposed to onslaughts from other life-forms, particularly small ones like mosquitoes. On one sultry day in the summer of 1879, near

Bergen Point, New Jersey, a furious swarm of these pests interrupted his study of some fine specimens of false indigo. Quickly plucking a branch from a smooth sumac that grew nearby, he waved it vigorously to drive off the hungry swarm and continued to examine the indigo. Sweaty work! When it was done, he found refreshment by popping some red-ripe sumac berries in his mouth and savoring their tartness. On return home after this episode—but let Dr. Millspaugh tell of his adventure in his own words. The exclamation point is his.

> For three successive nights following this occurrence I flew (!) over the city of New York with a graceful and delicious motion that I would give several years of my life to experience in reality. Query: Did I absorb from my perspiring hands sufficient juice of the [sumac] bark to produce the effect of the drug, or was it from the berries I held in my mouth?

He did not fly again. Nor, that I know of, has his flight been replicated by anyone since.

What *can* be replicated, and with the greatest ease, is his enjoyment of the berries' taste. Sumac yields a drink as good as any beverage —iced tea, cola, Gatorade—ever devised to quench a thirst. Drinking it down was once thought to be a sure way to reduce a fever, a belief reflected in staghorn sumac's scientific name, R. *typhina*—"fever" sumac. But taste is what really counts—the tart, lemony freshness, created by the malic acid (found also in apples) that clings to the fine crimson hairs thickly covering each berry. To make the drink, simply gather the ripe berry clusters in summer or early fall, when they've turned dark red and are still firm. As long as the fruit is the go-ahead red kind, the species of sumac doesn't matter. I use the one most common locally, the shiny sumac, *R. copallina*—"resin" sumac, for the sticky juice—which is also called dwarf, mountain, or flameleaf. (It should be this sumac that grants the power to fly: the stem between each set of pinnate leaves bears thin, translucent green wings.) Place berries to the halfway mark in a gallon pitcher, add cool water to the top, and refrigerate the whole works overnight. In the morning, remove the fruit by decanting the liquid through cheesecloth. Though you might expect otherwise, the resulting berryade is not red or pink but completely clear. And it's delicious by itself or with a suspicion—no more—of sweetener added.

But the deliciousness of red sumac berries, Dr. Millspaugh's remarkable flight, *Rhus* as a poison, and *Rhus* as a cure—these and all the other bits of sumac lore I've given here are only tangential moments in the sumac's tale. They pertain to humankind's perceptions of *Rhus*, not to the genus itself. The lore belongs to our stories, not those of a wonderfully diverse and peculiar plant.

Sumac, no matter what the species, is not fussy. In one guise or another, it grows with white pines, Virginia pines or loblollies; with oaks and maples, tulip trees, sweet gums, and persimmons; with exotics like paulownia and ailanthus that have snuggled themselves ineradicably into the New World's countryside. It grows in high places and low, along roadsides, in old fields and hedgerows, at the edges of woods. Smooth sumac, the marvel that once induced flight, is found in all forty-eight contiguous states. Nor does North American sumac of any kind confine itself to strictly rural areas but may also sprout in cities, where it shoots up, blooms, and flaunts autumn colors—yellow, gold, orange, scarlet—as boldly as any country cousin. Later, after the leaves are shed, the trunk of a young sumac may resemble nothing so much as a gawky stick with oval scars where leaf stems once clung. Older specimens spread cautious branches; their crowns are open and flat. The smooth sumac's young bark gives the species its name, for it is smooth indeed, in contrast to the twigs of the staghorn and shiny sumacs, both of which are coated with hairs. And these hairs, soft and brown as the velvet on a deer's developing antlers, account for staghorn's common name. When spring stimulates bloom, the small white flowers come in clusters that are either male or female, and only rarely do both kinds appear on the same tree. But reproduction is only sometimes a sexual affair, only sometimes a matter of pollination and seed production; it may also, in the fashion of many shrubs and trees, take a solitary, subterranean form. Underground runners shoot forth like hidden branches, and from them, new plants thrust strongly upward, reaching for light. Each one is genetically identical to its progenitor; each will establish its own root system and attain independence. New growth rising from the old—what dark exuberance in this business! It accounts for sumac's habit of sprouting up in thickets. It also accounts for the difficulties encountered by gardeners wanting to clear away a sneakily invasive stand: chop and dig as they will, pieces of severed runners, taking on life of their own, send up suckers —a child sticking out a tongue, a maniacal driver giving the finger, a plant exploding silently, witlessly out of containment. Unruliness? In sumac's terms, there's no such thing. There's only an active insistence on its own green right to live.

In my terms, sumac entices the senses: the plush velvet of staghorn twigs; the clean tartness of the fruit; the leaves turning bright, hot colors as soon as the nights turn cool; the earthen scent of berries cooking in the kettle and the lanolin smell of simmering wool; the stormy grey dye that stained my hand for a week and that for a quarter of a century has stayed true in the yarn.

Looking back down that corridor of years, I no longer remember what species of sumac furnished berries for the dye. Perhaps I never knew, never bothered to know. My attentions were focused on daylilies then, and junipers for holding steep banks, and vegetables for bringing to the table. Sumac was simply sumac—an ornament to fields and woods where it belonged, a plague in my carefully cultivated gardens.

I do remember working the rug. The skeins of freshly dyed yarn were hung out to dry, then rolled into balls. The life-story design was laid out and drawn on a length of monk's cloth with a marking pen. After the cloth had been firmly tacked wrong side up onto a rug frame, I set a punch needle for the desired loop height, threaded in newly colored yarn, and started brodding, as hooking with such a needle is called. Punch through the fabric and pull the needle back: the loops form below out of sight, while stitches as short and neat as hyphens show on top. The rug grew slowly. Autumn yielded to winter, snow came, and Christmas, and children's birthdays, along with all the must-be-dones and want-to-dos of daily life. The leaves on the trees in the yard had fully unfurled their leaves before it was done.

And to this day it quietly relates its many stories of many lives. Like fairy tales and Mother Goose rhymes learned in childhood and handed on to newer generations, they lose nothing—if anything, they gain—with each retelling. It's taken me till now, however, to see that still another story—the grandest, yet least obtrusive—lies within the overall design: without the underlying fabric, the rug and its community of stories could not exist. But with a connective tissue of cloth, the yarn holds firm and the lambent colors cohere in sensible design. And just as the dyer's hand takes on the color of the dye, the rug is subdued to the nature of the cloth, its suppleness, the tightness of its weave.

Surely the same principle applies in the world beyond the rug. Plants and animals, water and rocks and moving air—our different stories and disparate beings are held together in a shared ground. To subdue or to be subdued? The trick is, of course, to understand that Earth's intricately woven fabric is common to us all.

BOOKS TO LEAF THROUGH

Further Reading

TREES AND OTHER PLANTS

Bailey, Liberty Hyde. *How Plants Get Their Names.* New York: Dover Publications, Inc., 1963. A classic work first published in 1933. The dean of American botany here discusses topics from the life of Linnaeus to classification and nomenclature in a happily readable fashion.

Peattie, Donald Culross. *A Natural History of Trees of Eastern and Central North America.* Boston: Houghton Mifflin Company, 1991. A pleasantly chatty, highly enthusiastic hodgepodge of information by a noted writer on natural history. The book first appeared in 1948, and now reappears in a paperback edition. It retains every bit of its original value. The short, nontechnical articles cover a host of topics, from botany and historical reports to the personal acquaintance of the author with the species under consideration. A companion volume, *A Natural History of Western Trees,* is also available in paper.

Rogers, Julia Ellen. *The Tree Book: A Popular Guide to a Knowledge of the Trees of North America and to Their Uses and Cultivation.* Garden City, N.Y.: Doubleday, Page & Company, 1913. Botanical information liberally adorned with anecdotes from the writer's personal experience. The appreciative freshness and vigor of the latter give this book its great charm. Long out of print, it's well worth getting through an interlibrary loan.

Sargent, Charles Sprague. *Manual of the Trees of North America.* 2 vols. New York: Dover Publications, Inc.,1961. Unabridged republication of the 1922 edition, and my favorite general guide to American trees. Each description is accompanied by a drawing detailed and clear enough to aid in indentification.

The all-inclusive entries also contain easy-to-read facts about each species' measurements, range, habits, and human uses, along with a glossary explaining the cut-and-dried botanical terms. Some of the binomials have been superseded since 1922, but an update is included.

THE HISTORICAL RECORD

Bartram, William. *Travels through North & South Carolina, Georgia, East & West Florida.* Mark van Doren, editor. New York: Dover Publications, Inc., 1955. Tales of travel as lively today as when they were first published in 1791. The book not only preserves this American-born botanist's observations on many plants but hands on his bird lore, snake stories, and firsthand accounts of Native Peoples like the Creeks and the "Siminoles."

Feduccia, Alan. *Catesby's Birds of Colonial America.* Chapel Hill: The University of North Carolina Press, 1985. Reproductions (mostly, alas, in black and white) of the English naturalist Mark Catesby's sometimes primitive, always minutely attentive drawings of birds and plants. His descriptions, first published between 1731 and 1743, are also provided, along with modern commentary.

Harriot, Thomas. *A Briefe and True Report of the New Found Land of Virginia.* New York: Dover Publications, Inc., 1972. Republication of the 1590 edition, complete with its engravings of Indians at work and play. The first three parts of the *Report*—on things to sell, things to eat, and things to know about, respectively—contain many of the earliest European references to the vegetable wonders of the New World.

Kalm, Peter. *Travels into North America.* New York: Dover Publications, Inc., 1987. Republication of the 1770 English translation of journals that first saw the light of print in

Sweden in 1753. Not just plants but geography, fossils, earthquakes, tornadoes, fur skins, metals, and people, of course—from Indians to centenarians—furnished focal points for his eager attention.

Lawson, John. *A New Voyage to Carolina.* Hugh Talmage Leffler, editor. Chapel Hill: The University of North Carolina Press, 1976. A carefully edited and indexed edition of a book that first appeared in England in 1709. Beginning with the account of a long trek that Lawson made through the Carolinas from the coast to the mountains and back again, it includes extensive notes on geography, natural history, and the Native inhabitants, along with intriguing tidbits on matters like the Indian remedy for pica (roast a rear-mouse—a bat, that is—and feed it to the dirt-eating child). Much of Lawson's material on the Indians was so admired by later writers that they picked it up and presented it as their very own.

SEX AMONG THE VEGETABLES

Darwin, Erasmus. *The Botanic Garden, Part II, Containing The Loves of the Plants.* Introduction by Donald H. Reiman. New York: Garland Publishing, Inc., 1978. Facsimile of the second edition, London: J. Nichols, 1790. The work is snoopy, yes, and voyeuristic, but also somehow happy and innocent (as well as immensely but readably learned). The fun had by Dr. Darwin, his giddy, show-off delight at putting the passions of plants into poems, sparkles as brightly today as it did two hundred years ago.